The Polyhedral Face of Coordination Chemistry

The Polyhedral Face of Coordination Chemistry

Editors

Antonio Bianchi
Matteo Savastano

Basel • Beijing • Wuhan • Barcelona • Belgrade • Novi Sad • Cluj • Manchester

Editors
Antonio Bianchi
University of Florence
Sesto Fiorentino
Italy

Matteo Savastano
University San Raffaele Roma
Rome
Italy

Editorial Office
MDPI
St. Alban-Anlage 66
4052 Basel, Switzerland

This is a reprint of articles from the Special Issue published online in the open access journal *Crystals* (ISSN 2073-4352) (available at: https://www.mdpi.com/journal/crystals/special_issues/2J2BV01564).

For citation purposes, cite each article independently as indicated on the article page online and as indicated below:

Lastname, A.A.; Lastname, B.B. Article Title. *Journal Name* **Year**, *Volume Number*, Page Range.

ISBN 978-3-7258-1275-2 (Hbk)
ISBN 978-3-7258-1276-9 (PDF)
doi.org/10.3390/books978-3-7258-1276-9

© 2024 by the authors. Articles in this book are Open Access and distributed under the Creative Commons Attribution (CC BY) license. The book as a whole is distributed by MDPI under the terms and conditions of the Creative Commons Attribution-NonCommercial-NoDerivs (CC BY-NC-ND) license.

Contents

About the Editors vii

Sarah Kuß, Erica Brendler and Jörg Wagler
Molecular Structures of the Pyridine-2-olates PhE(pyO)$_3$ (E = Si, Ge, Sn)—[4+3]-Coordination at Si, Ge vs. Heptacoordination at Sn
Reprinted from: *Crystals* **2022**, *12*, 1802, doi:10.3390/cryst12121802 1

Julia Torres, Javier González-Platas and Carlos Kremer
Lanthanide(III) Complexes with Thiodiacetato Ligand: Chemical Speciation, Synthesis, Crystal Structure, and Solid-State Luminescence
Reprinted from: *Crystals* **2023**, *13*, 56, doi:10.3390/cryst13010056 15

Claudia Caltagirone, Maria Carla Aragoni, Massimiliano Arca, Alexander John Blake, Francesco Demartin, Alessandra Garau, et al.
Functionalization and Coordination Effects on the Structural Chemistry of Pendant Arm Derivatives of 1,4,7-trithia-10-aza-cyclododecane ([12]aneNS$_3$)
Reprinted from: *Crystals* **2023**, *13*, 616, doi:10.3390/cryst13040616 28

Matteo Savastano, Carlotta Cappanni, Carla Bazzicalupi, Cristiana Lofrumento and Antonio Bianchi
Anion Coordination into Ligand Clefts
Reprinted from: *Crystals* **2023**, *13*, 823, doi:10.3390/cryst13050823 40

Fabio Santanni, Laura Chelazzi, Lorenzo Sorace, Grigore A. Timco and Roberta Sessoli
Structural and Magnetic Properties of the {Cr(pybd)$_3$[Cu(cyclen)]$_2$}(BF$_4$)$_4$ Heteronuclear Complex
Reprinted from: *Crystals* **2023**, *13*, 901, doi:10.3390/cryst13060901 56

Yuxuan Zhang, Zheng Wei and Evgeny V. Dikarev
Synthesis, Structure, and Characterizations of a Heterobimetallic Heptanuclear Complex [Pb$_2$Co$_5$(acac)$_{14}$]
Reprinted from: *Crystals* **2023**, *13*, 1089, doi:10.3390/cryst13071089 69

Salvador Blasco, Begoña Verdejo, María Paz Clares and Enrique García-España
Transition Metals Meet Scorpiand-like Ligands
Reprinted from: *Crystals* **2023**, *13*, 1338, doi:10.3390/cryst13091338 78

Pulleng Moleko-Boyce, Eric C. Hosten and Zenixole R. Tshentu
Sulfonato Complex Formation Rather than Sulfonate Binding in the Extraction of Base Metals with 2,2′-Biimidazole: Extraction and Complexation Studies
Reprinted from: *Crystals* **2023**, *13*, 1350, doi:10.3390/cryst13091350 91

Sebastián Martínez, Carlos Kremer, Javier González-Platas and Carolina Mendoza
New Polynuclear Coordination Compounds Based on 2–(Carboxyphenyl)iminodiacetate Anion: Synthesis and X-rays Crystal Structures
Reprinted from: *Crystals* **2023**, *13*, 1669, doi:10.3390/cryst13121669 107

Edi Topić, Vladimir Damjanović, Katarina Pičuljan and Mirta Rubčić
Dinuclear Molybdenum(VI) Complexes Based on Flexible Succinyl and Adipoyl Dihydrazones
Reprinted from: *Crystals* **2024**, *14*, 135, doi:10.3390/cryst14020135 121

About the Editors

Antonio Bianchi

Antonio Bianchi received his Ph.D. in Chemistry at University of Florence in Italy in 1989. He was employed by the same university first as a tenured Researcher, then as an Associate Professor, and from 2000 as Professor of General and Inorganic Chemistry. From 2006 to 2009 he served as the Head of the Department of Chemistry "Ugo Schiff" of the said university. His divers research interests encompass inorganic chemistry, supramolecular chemistry, coordination of anions, thermodynamics of coordination compounds, catalysis and green energy production.

Matteo Savastano

Matteo Savastano received his Ph.D. in chemical sciences (inorganic and supramolecular chemistry) in 2018, from the University of Florence, Italy, under the supervision of Prof. Antonio Bianchi. After stays at the University of Jaén (Spain) and at the University of Valencia (Spain), he remained as a postdoctoral research fellow at the Department of Chemistry "Ugo Schiff" in Florence for five more years. In 2023 he moved to the San Raffaele University, based in Rome, Italy, where he currently is associate professor of general and inorganic chemistry. His main interests are anion and cation coordination chemistry, supramolecular interactions, solution equilibria, crystal structures and polyiodides.

Article

Molecular Structures of the Pyridine-2-olates PhE(pyO)$_3$ (E = Si, Ge, Sn)—[4+3]-Coordination at Si, Ge vs. Heptacoordination at Sn

Sarah Kuß [1], Erica Brendler [2] and Jörg Wagler [1,*]

[1] Institut für Anorganische Chemie, TU Bergakademie Freiberg, D-09596 Freiberg, Germany
[2] Institut für Analytische Chemie, TU Bergakademie Freiberg, D-09596 Freiberg, Germany
* Correspondence: joerg.wagler@chemie.tu-freiberg.de; Tel.: +49-3731-39-4343

Abstract: The phenyltetrel pyridine-2-olates PhE(pyO)$_3$ (E = Si, Ge, Sn; pyO = pyridine-2-olate) were synthesized from the respective chlorides PhECl$_3$ and 2-hydroxypyridine (2-pyridone) with the aid of a sacrificial base (triethylamine). Their solid-state structures were determined by single-crystal X-ray diffraction. PhSi(pyO)$_3$ exhibits a three-fold capped tetrahedral Si coordination sphere ([4+3]-coordination, Si···N separations ca. 3.0 Å), in accordance with structures of previously reported silicon pyridine-2-olates. PhGe(pyO)$_3$ adopts a related [4+3]-coordination mode, which differs in terms of the tetrahedral faces capped by the pyridine N atoms. Additionally, shorter Ge···N separations (2.8–2.9 Å) indicate a trend toward tetrel hypercoordination. PhSn(pyO)$_3$ features heptacoordinate tin within a pentagonal bipyramidal Sn coordination sphere (Sn···N separations 2.2–2.4 Å). For the Si and Sn compounds, ^{29}Si and ^{119}Sn NMR spectroscopy indicates retention of their tetrel coordination number in chloroform solution.

Keywords: 2-hydroxypyridine; hypercoordination; ^{119}Sn solid-state NMR; tetrel; X-ray diffraction

---

Citation: Kuß, S.; Brendler, E.; Wagler, J. Molecular Structures of the Pyridine-2-olates PhE(pyO)$_3$ (E = Si, Ge, Sn)—[4+3]-Coordination at Si, Ge vs. Heptacoordination at Sn. *Crystals* **2022**, *12*, 1802. https://doi.org/10.3390/cryst12121802

Academic Editor: Alexander Y. Nazarenko

Received: 14 November 2022
Accepted: 3 December 2022
Published: 10 December 2022

Publisher's Note: MDPI stays neutral with regard to jurisdictional claims in published maps and institutional affiliations.

Copyright: © 2022 by the authors. Licensee MDPI, Basel, Switzerland. This article is an open access article distributed under the terms and conditions of the Creative Commons Attribution (CC BY) license (https://creativecommons.org/licenses/by/4.0/).

1. Introduction

In previous studies [1–3] it was shown that in silicon pyridine-2-olates the potentially bidentate pyO group (pyO = pyridine-2-olate) is essentially monodentate, bound to Si through an Si–O bond and capping tetrahedral faces of the Si coordination sphere by rather remote N···Si coordination (e.g., in compounds **I** [2] and **II** [3], Figure 1). In addition, it has been shown that pyridine-2-olates may serve as bridging ligands with their N atom coordinating to a transition metal (TM) atom, thus fostering the formation of heteronuclear Si···TM complexes with higher-coordinate Si atom, for example in compounds **III** and **IV** [2] and some others [2,4]. Whereas in pyO-TM-silyl-complexes with tetracoordinate Si atom and thus formally covalent TM–Si bond (such as **V** [5] and **VI** [6]) pyO ligand(s) may buttress this diatomic core and thus simply add some stability, in cases with formally dative bonding or weak TM···Si attraction the availability of dangling N-donor sites in silicon pyridine-2-olates may be a key toward their binding to transition metals. With the perspective of extending studies of pyO-bridged TM-tetrel-complexes to heavier congeners of Si, the mode(s) of pyO-coordination at the heavier congeners (Ge, Sn) may hint at their suitability as starting materials for pyO-bridged tetrel-TM complexes. So far, molecular structures (in terms of crystallographically proven configurations) of germanium pyridine-2-olates are unknown [7]. For Sn-pyO-compounds, two crystal structures have been reported for compounds with penta- [8] or penta- and hexacoordinate Sn [9] in which the pyridine-2-olate is bridging two Sn atoms (with Sn–O and Sn–N bonds). These examples, however, involve a distannamethane motif [8] or a combination of different Sn sites (SnBu$_2$ and SnBu$_3$ groups) [9]. Crystallographic studies of pyO coordination in rather simple organotin compounds are yet to be performed. In the current study, we present a systematic comparison of the solid-state structures of the series of related phenyltetrel compounds PhE(pyO)$_3$ (E = Si, Ge, Sn). Whereas our interest in these compounds mainly arises from aspects of coordination chemistry (coordination of bidentate ligands at the tetrel and as bridging ligands between tetrel and transition metal), the exploration of coordination compounds of heavier tetrels with pyridine bases may also be of interest for other research fields. Both pyridine bases as ligands [10,11] and tin as a central atom [12] are the focus of the development of anti-cancer drugs.

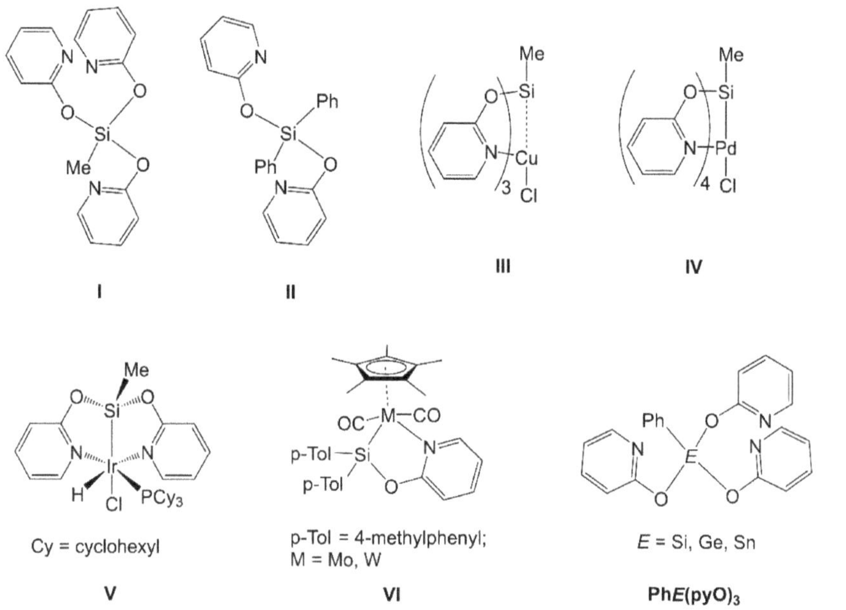

Figure 1. Selected pyridine-2-olate compounds of silicon with remote N···Si coordination (**I,II**), with hypercoordinate Si atom featuring a metal atom in the coordination sphere (**III,IV**) and with tetracoordinate Si bound to a transition metal atom (**V,VI**) as well as a generic representation of the compounds (**PhE(pyO)$_3$**) under investigation in this paper.

2. Materials and Methods

2.1. General Considerations

Starting materials 2-hydroxypyridine (ABCR, Karlsruhe, Germany, 98%), phenyltrichlorosilane (Wacker, Burghausen, Germany), phenyltrichlorogermane (Gelest, Tullytown, PA, USA, 95%) and phenyltrichlorostannane (Sigma-Aldrich, Steinheim, Germany, 98%) were used as received without further purification. THF, diethyl ether, and triethylamine were distilled from sodium/benzophenone and kept under argon atmosphere. Chloroform, stabilized with amylenes (Honeywell, Seelze, Germany, ≥99.5%) and CDCl$_3$ (Deutero, Kastellaun, Germany, 99.8%) were stored over activated molecular sieves (3 Å) for at least 7 days and used without further purification. All reactions were carried out under an atmosphere of dry argon utilizing standard Schlenk techniques. Solution NMR spectra (^{1}H, ^{13}C, ^{29}Si, ^{119}Sn) (cf. Figures S1–S9 in the supporting information) were recorded on Bruker Avance III 500 MHz and Bruker Nanobay 400 MHz spectrometers. ^{1}H, ^{13}C and ^{29}Si chemical shifts are reported relative to Me$_4$Si (0 ppm) as internal reference. ^{119}Sn chemical shifts are reported relative to Me$_4$Sn (0 ppm) with external referencing. ^{1}H and ^{13}C NMR signals were assigned in accord with mutual coupling patterns (in case of ^{1}H) and according to the shifts of corresponding ^{1}H or ^{13}C NMR signals in related compounds MeSi(pyO)$_3$ [2], Ph$_2$Si(pyO)$_2$ [3] and PhP(pyO)$_2$ [13]. Furthermore, ^{1}H–^{13}C HSQC techniques were employed for ^{13}C NMR signal assignment of compound PhSn(pyO)$_3$. The ^{119}Sn MAS NMR spectrum of PhSn(pyO)$_3$ was recorded on a Bruker Avance 400 WB spectrometer using a 4 mm zirconia (ZrO$_2$) rotor and an MAS frequency of v_{spin} = 13 kHz. The chemical shift is reported relative to Me$_4$Sn (0 ppm) and was referenced with the aid of a sample of SnO$_2$ (δ_{iso} = −603 ppm). Determination of the chemical shift anisotropy (CSA) tensor principal components from the spinning sideband spectrum was carried out with the SOLA module contained in the Bruker software package TOPSPIN. Principal components δ_{11}, δ_{22}, δ_{33} as well as span Ω and skew κ are reported according to the Herzfeld–Berger notation [14,15]. Elemental analyses were performed on an Elementar Vario MICRO cube. For single-crystal X-ray diffraction analyses, crystals were selected under an inert oil and mounted on a glass capillary (which was coated with silicone grease). Diffraction data were collected on a Stoe IPDS-2/2T diffractometer (STOE, Darmstadt, Germany) using Mo Kα-radiation. Data integration and absorption correction were performed with the STOE software XArea and XShape, respectively. The structures were solved by direct methods using SHELXS-97 or SHELXT and refined with the full-matrix least-squares methods of F^2 against all reflections with SHELXL-2014/7 [16–19]. All non-hydrogen atoms were anisotropically refined, and hydrogen atoms were isotropically refined in idealized position (riding model). For details of data collection and refinement (incl. the use of SQUEEZE in the refinement of the structure of PhSi(pyO)$_3$ · THF) see Appendix A, Table A1. Graphics of molecular structures were generated with ORTEP-3 [20,21] and POV-Ray 3.7 [22]. CCDC 2217508 (PhSi(pyO)$_3$ · THF), 2217510 (PhSi(pyO)$_3$ · CHCl$_3$), 2217509 (PhGe(pyO)$_3$), and 2217511 (PhSn(pyO)$_3$) contain the supplementary crystal data for this article. These data can be obtained free of charge from the Cambridge Crystallographic Data Centre via https://www.ccdc.cam.ac.uk/structures/ (accessed on 4 November 2022).

2.2. Syntheses and Characterization

Compound PhSi(pyO)$_3$ · THF (C$_{25}$H$_{25}$N$_3$O$_4$Si). A Schlenk flask was charged with magnetic stirring bar and 2-hydroxypyridine (2.00 g, 20.8 mmol), then evacuated and set under Ar atmosphere prior to adding THF (70 mL) and triethylamine (2.50 g, 24.7 mmol). The resultant mixture was stirred at room temperature, and phenyltrichlorosilane (1.60 g, 7.56 mmol) was added dropwise via syringe through a septum. Upon completed addition of silane, stirring was continued for 30 min, whereupon the flask was stored at 5 °C overnight. Thereafter, the triethylamine hydrochloride precipitate was removed by filtration and washed with THF (2 × 5 mL). From the combined filtrate and washings, the solvent was removed under reduced pressure (condensation into a cold trap) to afford a colorless solid. This crude product was dissolved into hot THF (4 mL) and allowed to crystallize upon cooling to room temperature. From this coarse crystalline product (thick colorless needles) of PhSi(pyO)$_3$···THF the supernatant was removed by decantation; the crystals were washed with THF (2 × 5 mL) and briefly dried in vacuum. Yield: 2.29 g (4.98 mmol, 66%). The yield is reported with respect to the composition PhSi(pyO)$_3$···THF, which is in accord with the composition of a single crystal taken from the freshly crystallized product for single-crystal X-ray diffraction analysis. ^{1}H NMR spectroscopy (cf. Figure S1 in the supporting information) already indicates some loss of THF upon drying. This effect was also found with elemental analysis: elemental analysis for C$_{21}$H$_{17}$N$_3$O$_3$Si···0.5 THF (423.52 g·mol^{-1}): C, 65.23%; H, 5.00%; N, 9.92%; found C, 65.21%; H, 5.10%; N, 9.96%. ^{1}H NMR (CDCl$_3$): δ (ppm) 8.11–8.07 (m, 2H, Ph-o), 8.01 (m, br, 3H, H^6), 7.52 (m, br, 3H, H^4), 7.44–7.32 (m, 3H, Ph-m/p), 6.92 (d, br, 3H, 8.2 Hz, H^3), 6.82 (m, br, 3H, H^5); ^{13}C{^{1}H} NMR (CDCl$_3$): δ (ppm) 160.3 (C^2), 147.3 (C^6), 139.1 (C^4), 135.6 (Ph-o), 130.9 (Ph-p), 129.0 (Ph-i), 127.6 (Ph-m), 118.2 (C^5), 113.0 (C^3); ^{29}Si{^{1}H} NMR (CDCl$_3$): δ (ppm) –64.7.

Some crystals of the chloroform solvate PhSi(pyO)$_3$ · CHCl$_3$ (C$_{22}$H$_{18}$Cl$_3$N$_3$O$_3$Si) were obtained by recrystallization of the THF solvate in chloroform.

Compound PhGe(pyO)$_3$ (C$_{21}$H$_{17}$GeN$_3$O$_3$). A Schlenk flask was charged with magnetic stirring bar and 2-hydroxypyridine (0.25 g, 2.63 mmol), then evacuated and set under Ar atmosphere prior to adding THF (5 mL) and triethylamine (0.40 g, 3.96 mmol). The resultant mixture was stirred at room temperature, and phenyltrichlorogermane (0.25 g, 0.97 mmol) was added dropwise via syringe through a septum. Upon completion, addition of the germane, a thick suspension was obtained. Therefore, further THF (2 mL) was added with stirring, before the flask was stored at 5 °C for five days. Thereafter, the triethylamine hydrochloride precipitate was removed by filtration and washed with THF (3 mL). From the combined filtrate and washings, the solvent was removed under reduced pressure (condensation into a cold trap) to afford a viscous oily residue. This crude product was dissolved into THF (0.5 mL), then diethyl ether (1 mL) was added at room temperature, and the solution was stored undisturbed at room temperature for crystallization to commence. In the course of some days, crystals of PhGe(pyO)$_3$ formed. For enhanced yield, the flask was stored at −24 °C overnight prior to isolation of the product, which was achieved by decantation of the supernatant, washing with diethyl ether (1 mL), and drying in vacuum. Yield: 0.26 g (0.60 mmol, 68%). A single crystal was taken from this product for single-crystal X-ray diffraction analysis. Elemental analysis for $C_{21}H_{17}GeN_3O_3$ (431.99 g·mol^{-1}): C, 58.39%; H, 3.97%; N, 9.73%; found C, 58.23%; H, 4.38%; N, 9.79%. ^{1}H NMR (CDCl$_3$): δ (ppm) 8.13–8.09 (m, 2H, Ph-o), 7.86 (m, 3H, H^6), 7.49 (m, 3H, H^4), 7.44–7.34 (m, 3H, Ph-m/p), 6.80 (d, 3H, 8.3 Hz, H^3), 6.70 (m, 3H, H^5); ^{13}C{^{1}H} NMR (CDCl$_3$): δ (ppm) 163.3 (C^2), 145.9 (C^6), 139.5 (C^4), 134.3 (Ph-o), 133.8 (Ph-i), 130.8 (Ph-p), 128.2 (Ph-m), 116.2 (C^5), 111.7 (C^3).

Compound PhSn(pyO)$_3$ ($C_{21}H_{17}N_3O_3Sn$). A Schlenk flask was charged with magnetic stirring bar and 2-hydroxypyridine (2.03 g, 21.4 mmol), then evacuated and set under Ar atmosphere prior to adding THF (40 mL) and triethylamine (2.66 g, 26.3 mmol). The resultant mixture was stirred in an ice/ethanol bath (ca. −10 °C), and phenyltrichlorostannane (2.18 g, 7.21 mmol) was added dropwise via syringe through a septum. Upon completed addition of the stannane, the mixture was stored at 5 °C for five days, whereupon the triethylamine hydrochloride precipitate was removed by filtration and washed with THF (10 mL). From the combined filtrate and washings, the solvent was removed under reduced pressure (condensation into a cold trap) to afford a white solid residue. This crude product was recrystallized from hot THF. The colorless solid product thus obtained was filtered off, washed with THF (2 mL) and dried in vacuum. Yield: 2.13 g (4.46 mmol, 63%). A single crystal was taken from this product for single-crystal X-ray diffraction analysis. Elemental analysis for $C_{21}H_{17}N_3O_3Sn$ (478.09 g·mol^{-1}): C, 52.76%; H, 3.58%; N, 8.79%; found C, 52.64%; H, 3.67%; N, 8.73%. ^{1}H NMR (CDCl$_3$): δ (ppm) 7.71–7.62 (m, 5H, Ph-o, H^6), 7.57 (m, 3H, H^4), 7.33–7.29 (m, 3H, Ph-m/p), 6.68 (d, 3H, 8.4 Hz, H^3), 6.58 (m, 3H, H^5); ^{13}C{^{1}H} NMR (CDCl$_3$): δ (ppm) 167.0 (22 Hz, C^2), 145.6 (1451 Hz, 1386 Hz, Ph-i), 142.3 (C^6), 141.8 (18 Hz, C^4), 133.7 (77 Hz, Ph-o), 129.5 (25 Hz, Ph-p), 128.6 (128 Hz, 122 Hz, Ph-m), 112.5 (C^5), 111.3 (36 Hz, C^3); ^{119}Sn{^{1}H} NMR (CDCl$_3$): δ (ppm) −609; ^{119}Sn CP/MAS NMR: δ_{iso} (ppm) −617.

Note: For some ^{13}C signals we observed $^{117/119}$Sn satellites, the ^{117}Sn and ^{119}Sn contributions of which are not resolved. Therefore, the average $J(^{117/119}$Sn-^{13}C) coupling constants are given in parentheses for those signals where applicable. For the pair of Ph-o and Ph-m ^{13}C NMR signals (at 133.7 and 128.6 ppm) the larger $J(^{117/119}$Sn-^{13}C) coupling is observed for the signal at 128.6 ppm. The assignment to o- and m-position is in accord with ^{13}C NMR data of other phenyltin compounds, e.g., PhSn(ESiMe$_2$)$_3$SiMe (E = S, Se, Te), which give rise to signals of Ph-o and Ph-m carbon atoms around 134–135 ppm and 128–129 ppm, respectively [23], and for Me$_3$SnPh the same phenomenon of ^3J(SnC) > ^2J(SnC) has been reported [24].

3. Results and Discussion

3.1. Crystallographic Analysis of the Molecular Structures of PhE(pyO)$_3$ (E = Si, Ge, Sn)

Compounds PhE(pyO)$_3$ (E = Si, Ge, Sn) were synthesized from the respective chlorides PhECl$_3$ and 2-hydroxypyridine in THF with the aid of triethylamine as a sacrificial base (Scheme 1). In all cases, the crystalline products obtained were of sufficient quality for single-crystal X-ray diffraction analysis (Table A1). The crystallographically determined molecular structures of PhE(pyO)$_3$ (E = Si, Ge, Sn) are shown in Figure 2 and selected interatomic separations and bond angles are given in Table 1. Whereas the Ge- and Sn-compound crystallized from THF without including solvent of crystallization, compound PhSi(pyO)$_3$ crystallized as a THF solvate PhSi(pyO)$_3$···THF with severely disordered solvent. Recrystallization from chloroform afforded an isomorphous solvate PhSi(pyO)$_3$ · CHCl$_3$, the disordered solvent of which could be refined in a satisfactory manner. As the structure models of PhSi(pyO)$_3$ in the two solvates are of similar quality, and the molecular conformation of the silane in those structures is the same, the data from the solvate PhSi(pyO)$_3$ · CHCl$_3$ will be used in the further discussion as a representative example.

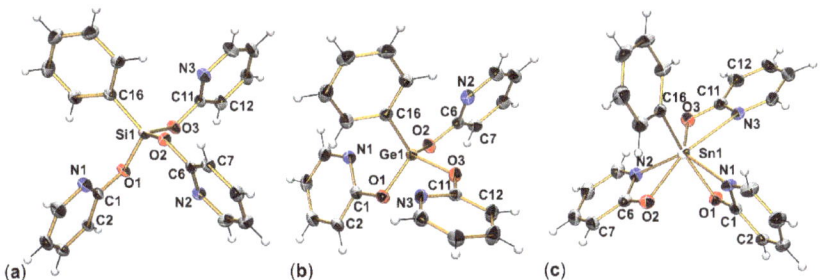

Scheme 1. Generic scheme of the syntheses of compounds PhSi(pyO)$_3$, PhGe(pyO)$_3$, and PhSn(pyO)$_3$. The reactions were performed at room temperature for PhSi(pyO)$_3$ and PhGe(pyO)$_3$ and at −10 °C for PhSn(pyO)$_3$.

Figure 2. Molecular structures of (**a**) PhSi(pyO)$_3$ (in the crystal structure of the solvate PhSi(pyO)$_3$·CHCl$_3$), (**b**) PhGe(pyO)$_3$, and (**c**) PhSn(pyO)$_3$ with thermal displacement ellipsoids at the 50% probability level and labels of selected non-hydrogen atoms. Selected interatomic distances (Å) and angles are listed in Table 1.

Table 1. Selected interatomic separations [Å] and bond angles [deg] in the tetrel (E = Si, Ge, Sn) coordination spheres of PhSi(pyO)$_3$, PhGe(pyO)$_3$ and PhSn(pyO)$_3$ and their pyO ligands.

Bond	PhSi(pyO)$_3$ · CHCl$_3$	PhGe(pyO)$_3$	PhSn(pyO)$_3$
E1–O1	1.640(1)	1.790(2)	2.120(2)
E1–O2	1.636(1)	1.777(2)	2.257(2)
E1–O3	1.642(1)	1.785(2)	2.133(2)
E1⋯N1	2.968(2)	2.789(2)	2.281(3)
E1⋯N2	3.002(2)	2.923(2)	2.212(3)
E1⋯N3	3.000(2)	2.908(2)	2.406(3)
E1–C16	1.842(2)	1.919(2)	2.138(2)
O1-E1-O2	115.85(6)	101.36(6)	84.64(9)
O1-E1-O3	96.91(6)	103.17(6)	93.00(9)
O2-E1-O3	112.54(6)	98.20(6)	138.85(9)
O1-C1-N1	117.50(13)	116.77(17)	113.0(3)
O2-C6-N2	117.30(13)	117.65(17)	113.9(3)
O3-C11-N3	117.45(13)	117.89(17)	113.5(3)
O1-C1-C2	117.55(13)	119.37(18)	126.2(3)
O2-C6-C7	117.96(13)	118.23(18)	126.7(3)
O3-C11-C12	117.47(13)	118.03(17)	124.3(3)

In compounds PhE(pyO)$_3$ E = Si, Ge the tetrel atom is essentially tetracoordinate, whereas in PhSn(pyO)$_3$ the tin atom is clearly heptacoordinate within a pentagonal bipyramidal coordination sphere. In the latter, O1 and C16 occupy the axial positions (O1-Sn1-C16 161.61(14)°), and the equatorial angles of pairs of neighboring bonds are 58.1(1)° (O3-Sn1-N3) and 59.2(1)° (O2-Sn1-N2) for the two chelate motifs and ca. 80° in the case of the remaining three angles spanned by O2,N1 (80.4(1)°); O3,N2 (79.7(1)°); N1,N3 (80.4(1)°). Nonetheless, the molecular metrics of PhGe(pyO)$_3$ already hint at a transition toward hypercoordination of the heavy tetrel atom. Whereas the E–O bond lengths increase in an expected manner with an increasing atomic radius of E, the E···N separations already undergo some significant shortening (by more than 0.1 Å on average) upon going from E = Si to Ge. Thus, with respect to the N atom's van der Waals radius of 1.55 Å and the tetrels' van der Waals radii of 2.10 (Si), 2.11 (Ge), and 2.17 Å (Sn) [25], the average E···N separations are 81.9%, 78.5%, and 61.8% of the sum of van der Waals radii for the Si, Ge, and Sn compound, respectively. Further evidence for pronounced E···N attraction upon going from E = Si to Ge can be found in the O-C-N angle deformation of all pyridine-2-olate groups. Whereas in the silicon compound the O-C-N angles and their corresponding O-C-C angles are very similar (both ca. 117.5°), the O-C-N angles in the Ge compound are slightly smaller than the corresponding O-C-C angles, thus indicating the pronounced attractive Ge···N interaction. At the Ge atom in compound PhGe(pyO)$_3$, the trend of bond angles adheres to VSEPR, i.e., within the GeC$_1$O$_3$ coordination sphere, all O-Ge-O angles are smaller than the tetrahedral angle. In sharp contrast, PhSi(pyO)$_3$ features one very small O-Si-O angle (O1-Si1-O3 96.91(6)°), whereas the other two O-Si-O angles (112.54(6) and 115.85(6)°) are noticeably wider than the tetrahedral angle. (Similar features are found in compound MeSi(pyO)$_3$ [2].) This phenomenon can be attributed to the molecular conformation of PhSi(pyO)$_3$ with respect to its enhanced coordination sphere (i.e., [4+3]-coordination). In the case of PhGe(pyO)$_3$, each of the pyridine N atoms is capping a tetrahedral face *trans* to a Ge–O bond and thus widens the O-Ge-C angles. The tetrahedral face *trans* to the Ge–C bond lacks this effect and thus allows for mutual shrinkage of the O-Ge-O angles (sum of O-Ge-O angles 302.7°). In the case of PhSi(pyO)$_3$, N2 is capping the tetrahedral face *trans* to the Si–C bond and thus widens the O-Si-O angles (sum of O-Si-O angles 325.3°). As the tetrahedral face *trans* to Si1–O2 is devoid of a remote donor atom (thus not exerting any additional widening to angle O1-Si1-O3), and the capping of tetrahedral faces by N1 and N3 enforces further widening of angles O1-Si1-O2 and O3-Si1-O2, respectively, bond angle O1-Si1-O3 becomes particularly narrow. Noteworthy, with respect to the overall deformation of the tetrahedral coordination sphere about the tetrel atom, the geometry parameter $\tau_{4'}$ [26] (Si: 0.91, Ge: 0.83) indicates particular deformation in the case of the Ge compound. It originates from a noticeably wide angle O2-Ge1-C16 (124.3(1)°). This feature arises from a C–H···N2 contact, in which a phenyl ortho-H atom interferes with pyridine atom N2 and thus competes with the capping of the tetrahedral face *trans* to Ge1–O1 by atom N2.

In addition to visualizing the conformational differences between the two [4+3]-coordinate tetrel compounds PhSi(pyO)$_3$ and PhGe(pyO)$_3$, the view in Figure 2 demonstrates the relationship of the conformation of PhGe(pyO)$_3$ and the conformation of the pentagonal bipyramidal Sn-coordination compound PhSn(pyO)$_3$. Scheme 2 illustrates this hypothetical transition, starting from the molecular conformation of PhGe(pyO)$_3$ (Scheme 2a). In addition to the N atoms approaching the tetrel E, partial rotation of the pyO ligands indexed with "**2**" and "**3**" about the bonds shown with rotation arrows in Scheme 2a,b affords the molecular configuration of PhSn(pyO)$_3$ (Scheme 2c).

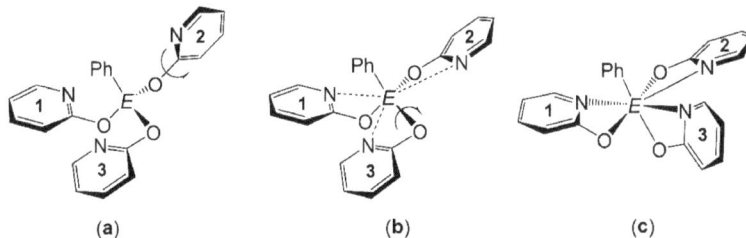

Scheme 2. Schematic representation of the ligand motion (of corresponding pyO moieties "**1**", "**2**" and "**3**") which relates the molecular conformations of PhGe(pyO)$_3$ (starting point (**a**)) and PhSn(pyO)$_3$ (represented by (**c**)) to one another via (**b**).

The increasing intensity of N⋯E coordination in the series E = Si < Ge < Sn is also reflected by the systematic changes in corresponding bond lengths within the pyO moieties (Table A2). Whereas most bond length differences merely allow for a vague hint at a trend (the changes, albeit seemingly systematic, are not significant within the boundaries of the standard deviations), significant shortening of the C–O bonds and, to a lesser extent, lengthening of the adjacent C–N bond is observed. Thus, the response of the pyO ligand´s C–C-bond backbone to mono- vs. bidentate coordination is less pronounced than the response of the related N-oxide (1-oxy-2-pyridinone, OPO) in compound (tBu)$_2$Si(OPO)$_2$, which features both a monodentate and a chelating OPO moiety within the same molecule [27].

The molecular conformation of PhSn(pyO)$_3$ requires some further discussion in context with the literature data. Even though this is the first crystallographically characterized "simple" stannane with more than one pyO substituent, some related compounds of the type RSn(pyS)$_3$ (pyS = pyridine-2-thiolate) have been reported and characterized crystallographically (with R = p-tolyl [28], Me and Ph [29]). In these compounds, the Sn atom is also heptacoordinate within an almost pentagonal bipyramidal coordination sphere, and the hydrocarbyl group as well as one chelate ligand´s chalcogen atom occupy axial positions. Their equatorial chelate ligands, however, are arranged in a *cis* fashion (thus giving rise to an equatorial N,N,N,S,S atom sequence), whereas in PhSn(pyO)$_3$ *trans*-arrangement of the two equatorial chelates is found (and an equatorial N,N,O,N,O atom sequence arises therefrom). With a different (O,N)-bidentate ligand system (ox = oxinate, 8-quinolinolate), compound RSn(ox)$_3$ (R = 4-chlorophenyl) with pentagonal bipyramidal Sn coordination sphere has been reported [30]. The conformation of this compound is less related to PhSn(pyO)$_3$ as it exhibits two differences: mutual *cis*-arrangement of the equatorial chelates and N-axial-O-equatorial arrangement of the third chelate ligand. Figure 3 illustrates this conformational difference.

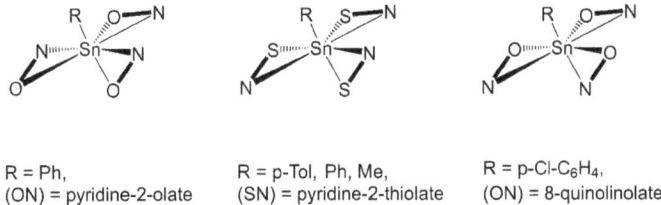

R = Ph,
(ON) = pyridine-2-olate

R = p-Tol, Ph, Me,
(SN) = pyridine-2-thiolate

R = p-Cl-C$_6$H$_4$,
(ON) = 8-quinolinolate

Figure 3. Generic illustration of the conformational difference between PhSn(pyO)$_3$, related pyridine-2-thiolates RSn(pyS)$_3$ [28,29] and 8-quinolinolate (4-Cl-C$_6$H$_4$)Sn(ox)$_3$ [30]. The motifs "N–O" and "N–S" serve as abbreviations of the different bidentate (N,O)- and (N,S)-donor ligands.

Apart from the comparison of PhSn(pyO)$_3$ with more or less related organotin-tris(chelates) with pentagonal bipyramidal Sn coordination sphere, the context of (Si,Ge,Sn)-hypercoordination within related four-membered (E,O,C,N)-chelate rings formed by mono-anionic chelators and tetravalent tetrels must be addressed. A search in the Cambridge

Crystal Structure Database [7] yielded a very limited number of (Si,Ge,Sn)-compounds with hypercoordinate (i.e., coordination number greater than four) tetrel atoms and such monoanionic (O,N)-chelating ligands. In the case of silicon, the portfolio is limited to three different compounds with hexacoordinate Si atoms, **VII** [31], **VIII** [32], and **IX** [33] (Figure 4). Enhanced Lewis acidity of the Si atom (caused by one or combinations of the parameter(s) such as small rings, electron-withdrawing substituents, and the cationic nature of the complex) may be an essential prerequisite. No representatives of hypercoordinate Ge compounds were found. Even the portfolio of structurally characterized hypercoordinate tetravalent tin compounds with this structural motif is very limited. For tin coordination numbers 6, 7, and 8, no appropriate hits were encountered. For Sn coordination number 7 only one multinuclear tin complex was found which features an (Sn,O,C,N)-chelate within a greater system of charge-delocalized multidentate ligands [34]. Even with tin coordination number 5, only two representatives (compounds **X**) were encountered. Interestingly, their four-membered (Sn,O,C,N)-chelates are derived from the pyridine-2-olate system [35].

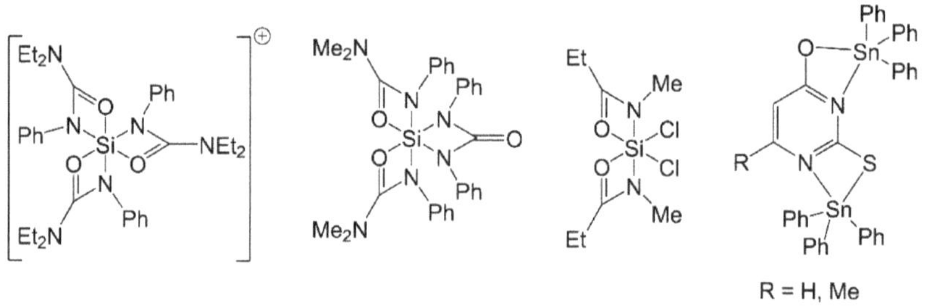

Figure 4. Literature known examples of structurally confirmed hypercoordinate tetravalent tetrel (E) compounds (compounds **VII–X**) with (E,O,C,N)-four-membered rings formed by mono-anionic (O,N)-chelating ligands.

In summary, structural characterization of hypercoordinate tetrels with four-membered (E,O,C,N)-chelates by mono-anionic chelators is still a rather unexplored field. With respect to pyridine-2-olate as a simple representative of such ligands, we may conclude that in the case of silicon compounds the lower tendency of the light tetrel toward hypercoordination can be seen as the major reason, as shown with PhSi(pyO)$_3$ and other examples [1–3]. In the case of Sn compounds, however, the lack of crystallographic evidence for solely Sn–O bound pyridine-2-olates with non-coordinating pyridine N atom [7] and the previously mentioned lack of structurally characterized Sn-pyO-chelates is highlighting a field of tetrel coordination chemistry yet to be explored.

3.2. NMR Spectroscopic Analyses of PhE(pyO)$_3$ (E = Si, Ge, Sn)

Compounds PhE(pyO)$_3$ (E = Si, Ge, Sn) were characterized by NMR spectroscopy in CDCl$_3$ solution and, in the case of PhSn(pyO)$_3$, with ^{119}Sn solid-state NMR spectroscopy.

In the CDCl$_3$ solution, the Si atom of compound PhSi(pyO)$_3$ is tetracoordinate, indicated by δ ^{29}Si = -64.7 ppm. This signal is upfield shifted with respect to that of the corresponding methyl compound MeSi(pyO)$_3$ (δ ^{29}Si = -46.5 ppm [2]), which is a common observation with pairs of corresponding PhSi and MeSi compounds (e.g., PhSi(OEt)$_3$ δ ^{29}Si = -57.8 ppm [36], MeSi(OEt)$_3$ δ ^{29}Si = -44.5 ppm [37]). In principle, compound PhSi(pyO)$_3$ has a similar ^{29}Si NMR shift as the tetracoordinate phenyl silicon compound PhSi(OEt)$_3$.

For compound PhSn(pyO)$_3$ the ^{119}Sn NMR shifts were recorded in the solid state (-617 ppm, Figure 5) as well as in CDCl$_3$ at different temperatures (20 °C: -609 ppm, -40 °C: -614 ppm). These chemical shifts are basically speaking for retention of the tin coordination number 7 in the CDCl$_3$ solution. Furthermore, these shifts are similar to those of other stannanes with heptacoordinate Sn atom and ArylSn(O,N)$_3$ coordination, e.g., p-TolSn(quinoline-8-olate)$_3$ (−611 ppm), p-TolSn(pyridine-2-carboxylate)$_3$ (−620 ppm) [30].

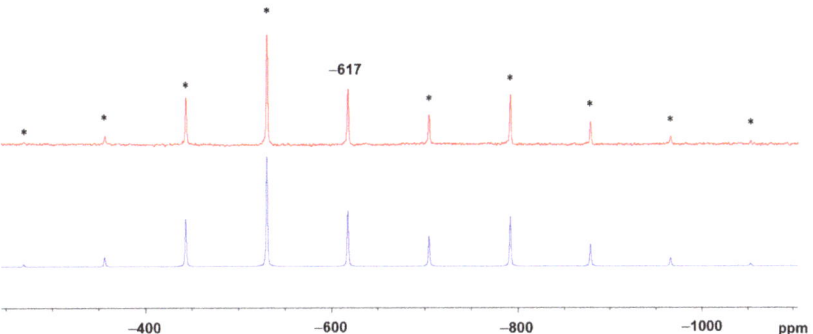

Figure 5. ^{29}Si MAS NMR spectrum of PhSn(pyO)$_3$ recorded at a spinning frequency of 13 kHz (top, red trace) and simulated spectrum using the CSA tensor data derived from the experimental spectrum (bottom, blue trace). The asterisked (*) peaks are spinning sidebands. CSA tensor data: δ_{iso} −617.4 ppm, δ_{11} −393.5 ppm, δ_{22} −502.2 ppm, δ_{33} −956.5 ppm, Ω 563 ppm, κ 0.61.

In the ^{1}H and ^{13}C NMR spectra of each of the compounds PhE(pyO)$_3$ (E = Si, Ge, Sn), the three pyO moieties give rise to one set of four (^{1}H) or five (^{13}C) signals, which was expected at least for the Si compound in accord with previously reported pyO-functionalized silanes [2,3]. In the case of compound PhSn(pyO)$_3$, it confirms that the SnCN$_3$O$_3$ coordination sphere itself is highly flexible and undergoes rapid exchange processes. Even cooling to −40 °C did not cause any signal broadening or decoalescence effects (Figure 6).

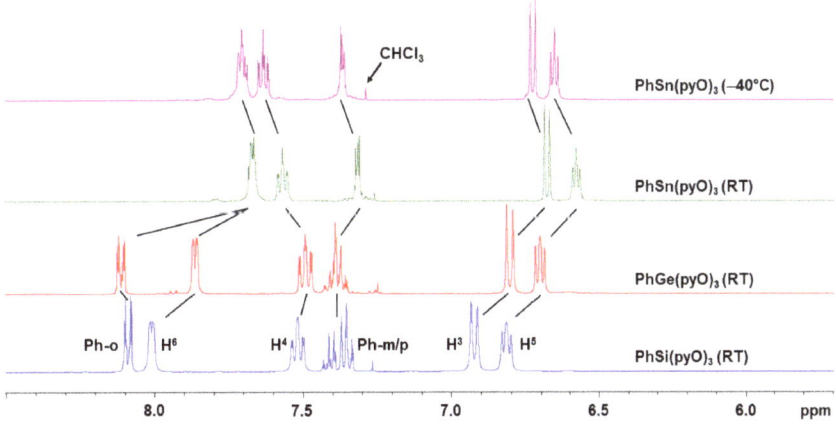

Figure 6. Section of the ^{1}H NMR spectra of CDCl$_3$ solutions of (from bottom to top) PhSi(pyO)$_3$, PhGe(pyO)$_3$, and PhSn(pyO)$_3$ at room temperature as well as of PhSn(pyO)$_3$ at −40 °C. The spectra are internally referenced to SiMe$_4$ (cf. spectra shown in Figures S1, S4, and S6 in the supplementary material).

To evaluate the pyO@Ge coordination chemistry of PhGe(pyO)$_3$ in CDCl$_3$ solution, a comparison of ^{1}H and ^{13}C NMR spectra of this compound with those of the lighter and heavier congener was engaged. In ^{13}C NMR spectra, it was found that the C^2 and C^6 positions of the pyO groups are highly responsive to coordinative changes at the N atom. In the spectrum of a CDCl$_3$ solution of compound PhP(pyO)$_2$ [13] the signals of interest emerge at 161.8 ppm (C^2) and 147.4 ppm (C^6). The spectrum of the related compound PhSb(pyO)$_2$ [38], which exhibits pronounced Sb⋯N attraction, exhibits the corresponding signals at 167.0 ppm (C^2) and 140.6 ppm (C^6). The corresponding signals of the related arsenic compound PhAs(pyO)$_2$ [39] emerge at intermediate positions. Even though this trend may also be influenced by electronic effects through the E–O–C$_5$N σ- and π-bond system (originating from different E–O bonds caused by the different elements E), the relevance of the electronic effects caused by N→tetrel coordination is underlined by corresponding data of a set of literature known pyridine-2-thiolate (pyS) compounds [3,40,41], which show that the same trends are observed for C^2 and C^6, but these trends are more responsive to hypercoordination (E(S,N)-chelation vs. absence of E···N coordination) rather than to the element E itself (cf. Table A3).

The ^{1}H and ^{13}C NMR spectra of compound PhGe(pyO)$_3$ in CDCl$_3$ solution resemble both the signal patterns (^{1}H coupling patterns) and the chemical shifts of corresponding signals (both ^{1}H and ^{13}C) of the related silicon compound. For a comparison of their ^{1}H NMR signals, see Figure 6. For the series of spectra of PhE(pyO)$_3$ (E = Si, Ge, Sn) recorded at room temperature, systematic shift trends are observed for corresponding ^{1}H signals of the pyO moieties. The upfield shift of the ^{1}H NMR signal of the pyO protons in position 6 from E = Si via Ge to Sn is the most prominent trend among their pyO ^{1}H signals and thus underlines the changes in N···E coordination, which can be derived from ^{13}C NMR data (Table 2). A trend toward hypercoordination of the Ge atom is indicated by the ^{13}C NMR shifts of the pyO carbon atoms in positions 2 and 6. Their signals are further downfield shifted (C^2, by 3 ppm) and upfield shifted (C^6, by 1.4 ppm) in the Ge compound.

Table 2. ^{13}C NMR shifts of C^2 and C^6 of the pyridine-2-olate (pyO) ligands of compounds PhE(pyO)$_3$ in CDCl$_3$.

Compound	E···N(pyS) Coordination? [1]	$\delta\ ^{13}C(C^2)$	$\delta\ ^{13}C(C^6)$
PhSi(pyO)$_3$	0	160.3	147.3
PhGe(pyO)$_3$	intermediate	163.3	145.9
PhSn(pyO)$_3$	1	167.0	141.8

[1] According to the number of pyO groups in the molecule and the coordination number of E, all pyO groups (1) or no pyO groups (0) establish E···N coordination. The ^{13}C NMR shift trends indicate intermediate E···N coordination for the Ge compound.

4. Conclusions

In general, compounds such as the herein presented set PhE(pyO)$_3$ represent potential starting materials for the syntheses of heteronuclear complexes. For compounds such as MeSi(pyO)$_3$ [2], Si(7-azaindolyl)$_4$ [42], ClSi(mt)$_3$ [43], Si(mt)$_4$ [44], FPhE(o-C$_6$H$_4$-P^{i}Pr$_2$)$_2$ (E = Si, Sn) [45], FE(o-C$_6$H$_4$-PPh$_2$)$_3$ (E = Si, Ge, Sn) [46], FSi(CH$_2$CH$_2$PMe$_2$)$_3$ [47], MeSi(OCH$_2$PMe$_2$)$_2$(CH$_2$CH$_2$PMe$_2$) [48] it has been shown that their dangling donor atoms (N, S, P) may bind to transition metals and, in some cases, foster establishing of formally dative bonding of the transition metal to the tetrel. As dangling donor arms of their potentially bidentate substituents may be essential for binding to another metal atom (because of the absence of other reactive groups within the molecule), compounds PhSi(pyO)$_3$ and PhGe(pyO)$_3$ with their absent or poor E···N coordination appear more suitable for that purpose than the related tin compound, in which the N atoms are engaged in a competing situation, i.e., Sn···N coordination also in solution. The flexibility/mobility of the Sn coordination sphere, however, may still render compounds such as PhSn(pyO)$_3$ suitable for the same purpose. Detailed studies of the ligand qualities of Ge- and Sn-pyO-compounds are yet to be performed. For compounds of the type RSi(pyO)$_3$ we had already confirmed the suitability as a ligand for Cu(I) in the case of R = Me [2], and a study of related silanes (R = Ph, Benzyl, Allyl) is currently underway.

Supplementary Materials: The following supporting information can be downloaded at: https://www.mdpi.com/article/10.3390/cryst12121802/s1: NMR spectra (^{1}H, ^{13}C{^{1}H} and, where applicable, ^{29}Si{^{1}H} or ^{119}Sn{^{1}H}) of CDCl$_3$ solutions of compounds PhSi(pyO)$_3$ (Figures S1–S3), PhGe(pyO)$_3$ (Figures S4, S5), and PhSn(pyO)$_3$ (Figures S6–S9).

Author Contributions: Conceptualization, J.W.; investigation, S.K., E.B., and J.W.; writing—original draft preparation, J.W.; writing—review and editing, E.B. and J.W.; visualization, J.W.; supervision, J.W. All authors have read and agreed to the published version of the manuscript.

Funding: This research received no external funding.

Institutional Review Board Statement: Not applicable.

Informed Consent Statement: Not applicable.

Data Availability Statement: Not applicable.

Acknowledgments: The authors are grateful to Beate Kutzner, Franziska Gründler, and Mareike Weigel (TU Bergakademie Freiberg, Institut für Anorganische Chemie) for solution NMR (B.K.) and elemental microanalysis service (F.G., M.W.), and to the students Alexandra Becker and Arthur Hentschel (TU Bergakademie Freiberg) for preparing additional batches of compound PhSi(pyO)$_3$ · THF.

Conflicts of Interest: The authors declare no conflict of interest.

Appendix A

Table A1. Crystallographic data from data collection and refinement for PhSi(pyO)$_3$ (its THF solvate, and chloroform solvate), PhGe(pyO)$_3$, and PhSn(pyO)$_3$.

Parameter	PhSi(pyO)$_3$ · THF [1]	PhSi(pyO)$_3$ · CHCl$_3$ [2]	PhGe(pyO)$_3$	PhSn(pyO)$_3$ [3]
Formula	C$_{25}$H$_{25}$N$_3$O$_4$Si	C$_{22}$H$_{18}$Cl$_3$N$_3$S$_3$Si	C$_{21}$H$_{17}$GeN$_3$O$_3$	C$_{21}$H$_{17}$N$_3$O$_3$Sn
M_r	459.57	506.83	431.96	478.07
T (K)	200(2)	180(2)	180(2)	150(2)
λ (Å)	0.71073	0.71073	0.71073	0.71073
Crystal system	monoclinic	monoclinic	triclinic	monoclinic
Space group	$P2_1/c$	$P2_1/c$	$P\bar{1}$	$P2_1$
a (Å)	12.1634(6)	12.2304(4)	9.7609(3)	8.4129(3)
b (Å)	17.0246(10)	16.8989(4)	9.7775(3)	10.9605(2)
c (Å)	11.4268(7)	11.3681(3)	9.8668(3)	10.8580(3)
α (°)	90	90	82.957(2)	90
β (°)	101.450(4)	101.160(2)	83.434(2)	98.385(2)
γ (°)	90	90	86.564(2)	90
V (Å^3)	2319.1(2)	2395.13(11)	927.42(5)	990.51(5)
Z	4	4	2	2
ρ_{calc} (g·cm^{-1})	1.32	1.46	1.55	1.60
$\mu_{MoK\alpha}$ (mm^{-1})	0.1	0.5	1.7	1.3
$F(000)$	968	1040	440	476
θ_{max} (°), R_{int}	28.0, 0.0243	28.0, 0.0454	28.0, 0.0428	29.0, 0.0270
Completeness	99.8%	99.9%	100%	99.9%
Reflns collected	15,889	26,422	24,367	56,321
Reflns unique	5587	5563	4480	5258
Restraints	0	73	0	1
Parameters	253	353	254	254
GoF	1.072	1.053	1.077	1.256
R1, wR2 [$I > 2\sigma(I)$]	0.0425, 0.1078	0.0420, 0.0983	0.0310, 0.0687	0.0184, 0.0416
R1, wR2 (all data)	0.0639, 0.1203	0.0570, 0.1059	0.0368, 0.0708	0.0203, 0.0433
Largest peak/hole (e·Å^{-3})	0.23, −0.39	0.30, −0.34	0.30, −0.50	0.93, −0.96

[1] The solvent molecule (THF) was severely disordered over many sites and could not be refined in a satisfactory manner. Therefore, the solvent was treated with SQUEEZE as implemented in PLATON [49–51]. This procedure detected per unit cell, solvent accessible volume of 562 Å^3, and contributions of 160 electrons therein (well in accord with 160 electrons for the four THF molecules per unit cell, which have been omitted from refinement). [2] The solvent molecule (chloroform) was refined disordered over three sites with site occupancy factors 0.269(6), 0.316(4) and 0.415(6). [3] The absolute structure parameter χ_{Flack} of this non-centrosymmetric structure refined to −0.035(6).

Table A2. Bond lengths [Å] of the pyridine-2-olate moieties in PhSi(pyO)$_3$ · CHCl$_3$, PhGe(pyO)$_3$ and PhSn(pyO)$_3$.

Bond [1]	PhSi(pyO)$_3$ · CHCl$_3$	PhGe(pyO)$_3$	PhSn(pyO)$_3$
O1–C1	1.362(2)	1.350(2)	1.314(4)
O2–C6	1.364(2)	1.350(2)	1.288(4)
O3–C11	1.360(2)	1.350(2)	1.315(4)
N1–C1	1.322(2)	1.327(3)	1.347(4)
N2–C6	1.321(2)	1.324(3)	1.348(4)
N3–C11	1.319(2)	1.324(3)	1.337(4)
C1–C2	1.386(2)	1.391(3)	1.408(4)
C5–C6	1.387(2)	1.389(3)	1.421(4)
C11–C12	1.392(2)	1.391(3)	1.403(4)
C2–C3	1.379(2)	1.379(3)	1.375(5)
C7–C8	1.379(2)	1.376(3)	1.368(5)
C12–C13	1.379(2)	1.374(3)	1.380(6)
C3–C4	1.385(3)	1.387(3)	1.399(5)
C8–C9	1.386(3)	1.388(3)	1.394(6)
C13–C14	1.384(3)	1.389(3)	1.391(6)
C4–C5	1.372(3)	1.376(3)	1.370(5)
C9–C10	1.371(3)	1.375(3)	1.379(5)
C14–C15	1.374(3)	1.372(3)	1.378(6)
N1–C5	1.345(2)	1.343(3)	1.341(4)
N2–C10	1.349(2)	1.349(3)	1.347(4)
N3–C15	1.347(2)	1.347(3)	1.353(4)

[1] For clarity of presentation, groups of corresponding bonds within the set of individual ligands of each molecule were summarized within a block of the same shading.

Table A3. ^{13}C NMR shifts of C^2 and C^6 of the pyridine-2-thiolate (pyS) ligands of a series of related silicon and tin compounds in CDCl$_3$ or CD$_2$Cl$_2$ solution.

Compound	E	Solvent	E⋯N(pyS) Coordination? [1]	δ^{13}C(C2)	δ^{13}C(C6)
PhClSi(pyS)$_2$ [40]	Si	CD$_2$Cl$_2$	1	167.5	140.8
MeClSi(pyS)$_2$ [40]	Si	CD$_2$Cl$_2$	1	168.1	140.4
Si(pyS)$_4$ [41]	Si	CDCl$_3$	0.5	163.4	144.7
Sn(pyS)$_4$ [41]	Sn	CDCl$_3$	0.5	162.9	145.7
Ph$_2$Si(pyS)$_2$ [3]	Si	CDCl$_3$	0	158.1	147.5
Me$_2$Si(pyS)$_2$ [3]	Si	CDCl$_3$	0	158.7	148.8

[1] According to the number of pyS groups in the molecule and the coordination number of E, all pyS groups (1) or no pyS groups (0) establish E⋯N coordination. In the case of compounds E(pyS)$_4$, the coordination number of E is 6. In these compounds, two E⋯N coordinating and two non- E⋯N-coordinating pyS groups exchange in a dynamic equilibrium (thus, a formal contribution of 0.5 of each pyS group to the overall E⋯N coordination arises therefrom).

References

1. Kazimierczuk, K.; Dołęga, A. Department of A. Synthesis and structural characterization of new cyclic siloxane with functionalized organic substituents. *Phosphorus Sulfur Silicon Relat. Elem.* **2017**, *192*, 1140–1143. [CrossRef]
2. Ehrlich, L.; Gericke, R.; Brendler, E.; Wagler, J. (2-Pyridyloxy)silanes as Ligands in Transition Metal Coordination Chemistry. *Inorganics* **2018**, *6*, 119. [CrossRef]
3. Seidel, A.; Weigel, M.; Ehrlich, L.; Gericke, R.; Brendler, E.; Wagler, J. Molecular Structures of the Silicon Pyridine-2-(thi)olates Me$_3$Si(pyX), Me$_2$Si(pyX)$_2$ and Ph$_2$Si(pyX)$_2$ (py = 2-Pyridyl, X = O, S), and Their Intra- and Intermolecular Ligand Exchange in Solution. *Crystals* **2022**, *12*, 1054. [CrossRef]
4. Sun, J.; Ou, C.; Wang, C.; Uchiyama, M.; Deng, L. Silane-Functionalized N-Heterocyclic Carbene–Cobalt Complexes Containing a Five-Coordinate Silicon with a Covalent Co−Si Bond. *Organometallics* **2015**, *34*, 1546–1551. [CrossRef]
5. Julián, A.; Garcés, K.; Lalrempuia, R.; Jaseer, E.A.; García-Orduña, P.; Fernández-Alvarez, F.J.; Lahoz, F.J.; Oro, L.A. Reactivity of Ir−NSiN Complexes: Ir-Catalyzed Dehydrogenative Silylation of Carboxylic Acids. *ChemCatChem* **2018**, *10*, 1027–1034. [CrossRef]
6. Kanno, Y.; Komuro, T.; Tobita, H. Direct Conversion of a Si−C(aryl) Bond to Si−Heteroatom Bonds in the Reactions of η^3-α-Silabenzyl Molybdenum and Tungsten Complexes with 2-Substituted Pyridines. *Organometallics* **2015**, *34*, 3699–3705. [CrossRef]
7. This refers to a search in the Cambridge Structure Database using ConQuest version 2022.2.0.

8. Hadjikakou, K.; Jurkschat, K.; Schürmann, M. Novel organotin(IV) compounds derived from bis(organostannyl)methanes: Synthesis and crystal structures of bis[diphenyl(pyridin-2-onato)stannyl]methane and bis[bromophenyl(pyrimidine-2-thionato)stannyl]methane · C7H8. *J. Organomet. Chem.* **2006**, *691*, 1637–1642. [CrossRef]
9. Ma, C.; Li, Q.; Zhang, R. A novel self-assembling of mixed tri- and dibutyltin macrocyclic complex with solvothermal synthesis method. *Inorg. Chim. Acta* **2009**, *362*, 2937–2940. [CrossRef]
10. Haribabu, J.; Tamura, Y.; Yokoi, K.; Balachandran, C.; Umezawa, M.; Tsuchiya, K.; Yamada, Y.; Karvembu, R.; Aoki, S. Synthesis and Anticancer Properties of Bis- and Mono(cationic peptide) Hybrids of Cyclometalated Iridium(III) Complexes: Effect of the Number of Peptide Units on Anticancer Activity. *Eur. J. Inorg.Chem.* **2021**, *2021*, 1796–1814. [CrossRef]
11. Zhou, J.; Yu, G.; Yang, J.; Shi, B.; Ye, B.; Wang, M.; Huang, F.; Stang, P.J. Polymeric Nanoparticles Integrated from Discrete Organoplatinum(II) Metallacycle by Stepwise Post-assembly Polymerization for Synergistic Cancer Therapy. *Chem. Mater.* **2020**, *32*, 4564–4573. [CrossRef]
12. Yusof, E.N.M.; Ravoof, T.B.S.A.; Page, A.J. Cytotoxicity of Tin(IV)-based compounds: A review. *Polyhedron* **2021**, *198*, 115069. [CrossRef]
13. Gericke, R.; Wagler, J. Ruthenium complexes of phosphino derivatives of carboxylic amides: Synthesis and characterization of tridentate P,E$_2$ and tetradentate P,E$_3$ (E = N,O) ligands and their reactivity towards [RuCl$_2$(PPh$_3$)$_3$]. *Polyhedron* **2017**, *125*, 57–67. [CrossRef]
14. Herzfeld, J.; Berger, A.E. Sideband intensities in NMR spectra of samples spinning at the magic angles. *J. Chem. Phys.* **1980**, *73*, 6021–6030. [CrossRef]
15. Mason, J. Conventions for the reporting of nuclear magnetic shielding (or shift) tensors suggest by participants in the NATO AEW in NMR Shielding Constants at the University of Maryland, College Park, July 1992. *Solid State Nucl. Magn. Reson.* **1993**, *2*, 285–288. [CrossRef] [PubMed]
16. Sheldrick, G.M. *Program for the Solution of Crystal Structures*; SHELXS-97; University of Göttingen: Göttingen, Germany, 1997.
17. Sheldrick, G.M. SHELXT—Integrated space-group and crystal-structure determination. *Acta Crystallogr. A* **2015**, *71*, 3–8. [CrossRef] [PubMed]
18. Sheldrick, G.M. *Program for the Refinement of Crystal Structures*; SHELXL-2014/7; University of Göttingen: Göttingen, Germany, 2014.
19. Sheldrick, G.M. A short history of SHELX. *Acta Crystallogr. A* **2008**, *64*, 112–122. [CrossRef]
20. Farrugia, L.J. ORTEP-3 for windows—A version of ORTEP-III with a graphical user interface (GUI). *J. Appl. Crystallogr.* **1997**, *30*, 565. [CrossRef]
21. Farrugia, L.J. WinGX and ORTEP for Windows: An update. *J. Appl. Crystallogr.* **2012**, *45*, 849–854. [CrossRef]
22. POV-RAY (Version 3.7), Trademark of Persistence of Vision Raytracer Pty. Ltd.,Williamstown, Victoria (Australia). Copyright Hallam Oaks Pty. Ltd., 1994–2004. Available online: http://www.povray.org/download/ (accessed on 28 June 2021).
23. Herzog, U.; Rheinwald, G. Novel Chalcogenides of Silicon with Bicyclo[2.2.2]octane Skeletons, MeSi(SiMe$_2$E)$_3$MR (E = S, Se, Te; M = Si, Ge, Sn; R = Me, Ph). *Organometallics* **2001**, *20*, 5369–5374. [CrossRef]
24. Schaeffer, C.D., Jr.; Zuckerman, J.J. Pulsed fourier transform NMR of substituted aryltrimethyltin derivatives: II. (^{119}Sn-^{13}C) coupling constants and ^{13}C chemical shifts of meta- and para-derivatives. *J. Organomet. Chem.* **1973**, *55*, 97–110. [CrossRef]
25. Mantina, M.; Chamberlin, A.C.; Valero, R.; Cramer, C.J.; Truhlar, D.G. Consistent van der Waals Radii for the Whole Main Group. *J. Phys. Chem. A* **2009**, *113*, 5806–5812. [CrossRef] [PubMed]
26. Okuniewski, A.; Rosiak, D.; Chojnaki, J.; Becker, B. Coordination polymers and molecular structures among complexes of mercury(II) halides with selected 1-benzoylthioureas. *Polyhedron* **2015**, *90*, 47–57. [CrossRef]
27. Kraft, B.M.; Brennessel, W.W. Chelation and Stereodynamic Equilibria in Neutral Hypercoordinate Organosilicon Complexes of 1-Hydroxy-2-pyridinone. *Organometallics* **2014**, *33*, 158–171. [CrossRef]
28. Schürmann, M.; Huber, F. Tris(2-pyridinethiolato)(p-tolyl)tin(IV), [Sn(C$_5$H$_4$NS)$_3$(C$_7$H$_7$)]. *Acta Crystallogr. C* **1994**, *50*, 206–209. [CrossRef]
29. Huber, F.; Schmiedgen, R.; Schürmann, M.; Barbieri, R.; Ruisi, G.; Silvestri, A. Mono-organotin(IV) and Tin(IV) Derivatives of 2-Mercaptopyridine and 2-Mercaptopyrimidine: X-ray Structures of Methyl-tris(2-pyridinethiolato)tin(IV) and Phenyl-tris(2-pyridinethiolato)tin(IV)·1.5CHCl$_3$. *Appl. Organomet. Chem.* **1997**, *11*, 869–888. [CrossRef]
30. Schürmann, M.; Schmiedgen, R.; Huber, F.; Silvestri, A.; Ruisi, G.; Barbieri Paulsen, A.; Barbieri, R. Mono-aryltin(IV) and mono-benzyltin(IV) complexes with pyridine-2-carboxylic acid and 8-hydroxyquinoline. X-ray structure of *p*-chloro-phenyl-tris(8-quinolinato)tin(IV)·2CHCl$_3$. *J. Organomet. Chem.* **1999**, *584*, 103–117. [CrossRef]
31. Schöne, D.; Gerlach, D.; Wiltzsch, C.; Brendler, E.; Heine, T.; Kroke, E.; Wagler, J. A Distorted Trigonal Antiprismatic Cationic Silicon Complex with Ureato Ligands: Syntheses, Crystal Structures and Solid State ^{29}Si NMR Properties. *Eur. J. Inorg. Chem.* **2010**, *2010*, 461–467. [CrossRef]
32. Bohme, U. *CSD Private Communication* **2020**, CCDC 2025957.
33. Yang, J.; Verkade, J.G. Non-catalyzed addition reactions of Cl$_3$SiSiCl$_3$ with 1,2-diketones, 1,2-quinones and with a 1,4-quinone. *J. Organomet. Chem.* **2002**, *651*, 15–21. [CrossRef]
34. Feng, Y.-L.; Zhang, F.-X.; Kuang, D.-Z.; Yang, C.-L. Two Novel Dibutyltin Complexes with Trimers and Hexanuclear Based on the Bis(5-Cl/Me-salicylaldehyde) Carbohydrazide: Syntheses, Structures, Fluorescent Properties and Herbicidal Activity. *Chin. J. Struct. Chem.* **2020**, *39*, 682–692. [CrossRef]

35. Ma, C.; Tian, G.; Zhang, R. New triorganotin(IV) complexes of polyfunctional S,N,O-ligands: Supramolecular structures based on π···π and/or C–H···π interactions. *J. Organomet. Chem.* **2006**, *691*, 2014–2022. [CrossRef]
36. Boyer, J.; Brelière, C.; Carré, F.; Corriu, R.J.P.; Kpoton, A.; Poirier, M.; Royo, G.; Young, J.C. Five-co-ordinated Silicon Compounds: Geometry of Formation by Intramolecular Co-ordination. Crystal Structure of 2-(Dimethylaminomethyl)phenyl-I-naphthylsilane. *J. Chem. Soc. Dalton. Trans.* **1989**, *1*, 43–51. [CrossRef]
37. Engelhardt, G.; Radeglia, R.; Jancke, H.; Lippmaa, E.; Mägi, M. Zur Interpretation ^{29}Si-NMR-chemischer Verschiebungen. *Org. Magn. Reson.* **1973**, *5*, 561–566. [CrossRef]
38. Gericke, R.; Wagler, J. Ruthenium Complexes of Stibino Derivatives of Carboxylic Amides: Synthesis and Characterization of Bidentate Sb,E, Tridentate Sb,E$_2$, and Tetradentate Sb,E$_3$ (E = N and O) Ligands and Their Reactivity Toward [RuCl$_2$(PPh$_3$)$_3$]. *Inorg. Chem.* **2020**, *59*, 6359–6375. [CrossRef]
39. Gericke, R.; Wagler, J. (2-Pyridyloxy)arsines as ligands in transition metal chemistry: A stepwise As(III) → As(II) → As(I) reduction. *Dalton Trans.* **2020**, *49*, 10042–10051. [CrossRef] [PubMed]
40. Baus, J.A.; Burschka, C.; Bertermann, R.; Fonseca Guerra, C.; Bickelhaupt, F.M.; Tacke, R. Neutral Six-Coordinate and Cationic Five-Coordinate Silicon(IV) Complexes with Two Bidentate Monoanionic N,S-Pyridine-2-thiolato(−) Ligands. *Inorg. Chem.* **2013**, *52*, 10664–10676. [CrossRef]
41. Wächtler, E.; Gericke, R.; Kutter, S.; Brendler, E.; Wagler, J. Molecular structures of pyridinethiolato complexes of Sn(II), Sn(IV), Ge(IV), and Si(IV). *Main Group Met. Chem.* **2013**, *36*, 181–191. [CrossRef]
42. Wahlicht, S.; Brendler, E.; Heine, T.; Zhechkov, L.; Wagler, J. 7-Azaindol-1-yl(organo)silanes and Their PdCl$_2$ Complexes: Pd-Capped Tetrahedral Silicon Coordination Spheres and Paddlewheels with a Pd-Si Axis. *Organometallics* **2014**, *33*, 2479–2488. [CrossRef]
43. Autschbach, J.; Sutter, K.; Truflandier, L.A.; Brendler, E.; Wagler, J. Atomic Contributions from Spin-Orbit Coupling to ^{29}Si NMR Chemical Shifts in Metallasilatrane Complexes. *Chem. Eur. J.* **2012**, *18*, 12803–12813. [CrossRef]
44. Wagler, J.; Brendler, E. Metallasilatranes: Palladium(II) and Platinum(II) as Lone-Pair Donors to Silicon(IV). *Angew. Chem. Int. Ed.* **2010**, *49*, 624–627. [CrossRef]
45. Gualco, P.; Lin, T.-P.; Sircoglou, M.; Mercy, M.; Ladeira, S.; Bouhadir, G.; Pérez, L.M.; Amgoune, A.; Maron, L.; Gabbaï, F.P.; et al. Gold–Silane and Gold–Stannane Complexes: Saturated Molecules as σ-Acceptor Ligands. *Angew. Chem. Int. Ed.* **2009**, *48*, 9892–9895. [CrossRef] [PubMed]
46. Kameo, H.; Kawamoto, T.; Bourissou, D.; Sakaki, S.; Nakazawa, H. Evaluation of the σ-Donation from Group 11 Metals (Cu, Ag, Au) to Silane, Germane, and Stannane Based on the Experimental/Theoretical Systematic Approach. *Organometallics* **2015**, *34*, 1440–1448. [CrossRef]
47. Grobe, J.; Lütke-Brochtrup, K.; Krebs, B.; Läge, M.; Niemeyer, H.-H.; Würthwein, E.-U. Alternativ-Liganden XXXVIII. Neue Versuche zur Synthese von Pd(0)- und Pt(0)-Komplexen des Tripod-Phosphanliganden FSi(CH$_2$CH$_2$PMe$_2$)$_3$. *Z. Naturforsch.* **2007**, *62b*, 55–65. [CrossRef]
48. Grobe, J.; Krummen, N.; Wehmschulte, R.; Krebs, B.; Läge, M. Alternativ-Liganden. XXXI Nickelcarbonylkomplexe mit Tripod-Liganden des Typs XM'(OCH$_2$PMe$_2$)$_n$(CH$_2$CH$_2$PR$_2$)$_{3-n}$ (M' = Si, Ge; n = 0-3). *Z. Anorg. Allg. Chem.* **1994**, *620*, 1645–1658. [CrossRef]
49. Spek, A.L. Single-crystal structure validation with the program PLATON. *J. Appl. Cryst.* **2003**, *36*, 7–13. [CrossRef]
50. Spek, A.L. Structure validation in chemical crystallography. *Acta Crystallogr. D* **2009**, *65*, 148–155. [CrossRef] [PubMed]
51. Spek, A.L. PLATON SQUEEZE: A tool for the calculation of the disordered solvent contribution to the calculated structure factors. *Acta Crystallogr. C* **2015**, *71*, 9–18. [CrossRef]

Article

Lanthanide(III) Complexes with Thiodiacetato Ligand: Chemical Speciation, Synthesis, Crystal Structure, and Solid-State Luminescence

Julia Torres [1], Javier González-Platas [2] and Carlos Kremer [1],*

[1] Área Química Inorgánica, Departamento Estrella Campos, Facultad de Química, Universidad de la República, Montevideo 11800, Uruguay
[2] Departamento de Física, Instituto Universitario de Estudios Avanzados en Física Atómica, Molecular y Fotónica (IUDEA), MALTA-Cosolider Team, Universidad de La Laguna, E-38206 La Laguna, Tenerife, Spain
* Correspondence: ckremer@fq.edu.uy

Citation: Torres, J.; González-Platas, J.; Kremer, C. Lanthanide(III) Complexes with Thiodiacetato Ligand: Chemical Speciation, Synthesis, Crystal Structure, and Solid-State Luminescence. *Crystals* **2023**, *13*, 56. https://doi.org/10.3390/cryst13010056

Academic Editor: László Kovács

Received: 8 December 2022
Revised: 21 December 2022
Accepted: 24 December 2022
Published: 28 December 2022

Copyright: © 2022 by the authors. Licensee MDPI, Basel, Switzerland. This article is an open access article distributed under the terms and conditions of the Creative Commons Attribution (CC BY) license (https://creativecommons.org/licenses/by/4.0/).

Abstract: The synthesis, crystal structures, and luminescence of two lanthanide polynuclear complexes with the general formula $[Ln_2(tda)_3(H_2O)_5]\cdot 3H_2O$ (Ln = Sm, Eu; tda = thiodiacetato anion) are reported. The compounds were obtained by direct reaction of H_2tda and lanthanide(III) chloride in an aqueous solution. The choice of the conditions of synthesis was based on speciation studies. The structure of the polymeric complexes contains Ln(III) ions in a tricapped trigonal prism geometry. The versatility of this ligand provides different coordination modes and provokes the formation of thick 2D sheets. Direct excitation of the Ln(III) ions gives place to the characteristic intra-configuration sharp luminescence emission of both complexes in the solid state.

Keywords: lanthanide complexes; crystal structure; chemical speciation; thiodiacetato complex

1. Introduction

The study of lanthanide(III) (Ln) coordination compounds has elicited considerable interest in the last decades [1–4]. Different coordination numbers and geometries can give place to novel and interesting crystal structures. In addition, they have great potential as functional solid materials such as luminescent materials [5], magnetic devices [6], chemical sensors [7,8], etc. We and other groups have been interested in Ln(III) mononuclear and polynuclear complexes with ligands of type $X-(CH_2-COO)_2^{2-}$, where X = O (oxydiacetato, oda^{2-}), NH (iminodiacetato, ida^{2-}) or S (thiodiacetato, tda^{2-}) [9]. Oxydiacetato, the most deeply studied ligand of this group, appears as the most suitable ligand for Ln ions because of the presence of three O donor atoms in its structure. Hence, many complexes were already reported and characterized [9–13]. The thermodynamic stability of Ln-oda complexes also allows the use of the tris-chelate ($[Ln(oda)_3]^{3-}$) as a complex-as-ligand block towards M(II) cations [9,14]. The resulting heteropolynuclear compounds were assayed as catalysts [15,16], white-light emitters [17], and proton conductive MOF-based materials [18]. Iminodiacetato, with an N atom in the center, has also been studied, but to a lesser extent [9]. Substitution of O in oda by N in ida provokes a poorer participation of N in the coordination [14,19], which has hindered the isolation of tris-chelates.

Thiodiacetato is the less explored member of this series of ligands. The combination of O-carboxylate (hard donor atom) with sulfur (soft donor atom) makes the chemistry of this ligand very versatile and, at the same time, very challenging. Several structural reports show its capacity to act as a tridentate ligand towards Cu(II) [20–23], Ni(II) [24–29], Co(II) [30–36], Mn(II) [37], Zn(II) [38–40], Cd(II) [41], Mg(II) [42], Re(V) [43], V(IV) [44], and Ru(III) [45]. However, only one report can be found containing a bis-chelated fragment in a mononuclear compound [32], namely, $(pipH_2)[Co(tda)_2]\cdot 2H_2O$ ($pipH_2^{2+}$ = piperazine dication). The other structures always contain a coligand. The possibility of acting as a bidentate ligand (O,O or O,S) for these metal ions is restricted to a few cases [46,47]. Thiodiacetato ligand also exhibits the possibility of forming a bridge between metal ions, yielding polynuclear complexes. This can be found in structures with Cu(II) [20,48–50], Mn(II) [37,49,51], and Zn(II) [52–54].

Coordination compounds of tda with Ln ions are even less frequent. Anionic isolated tris-chelates have been reported in $(H_2Gun)_3[Ln(tda)_3]$ (Ln = Pr, Nd, HGun = guanidinium, $C(NH_2)_3^+$) [55]. In these structures, tda acts as bidentate through two O-carboxylate atoms, forming an 8-membered ring. S atom is 3.423 Å apart from Ln ion and does not participate in the coordination sphere. Another report presents a polynuclear compound with the formula $[Nd(tda)(H_2O)_4]Cl$ [56]. In this compound, tda acts as tridentate and additionally bridges Nd ions through carboxylate groups. A zigzag chain is formed. Nd(III) also forms an anionic 2D network in the complex $Na[Nd(tda)_2]$, in which the S atom is not coordinated, and the coordination sphere is filled only by O atoms from tda [57]. The structure of $(pipH_2)[Ce_2(tda)_4(H_2O)_2]\cdot 3H_2O$ is also a 2D anionic structure [58]. Finally, other groups of 2D polynuclear structures can be found. In $[Ln_2(tda)_3(H_2O)_2]_n$ (Ln = La, Sm, Gd, Nd, Pr, Tb, Dy, Eu), tda acts as bidentate and also as a ditopic ligand. The S atom seems to participate in the coordination but at a rather long distance [59–63].

In order to increase the knowledge of tda as a ligand towards Ln ions, we have revisited the solution chemistry of the systems by potentiometry and prepared, under mild reaction conditions, complexes with the general formula $[Ln_2(tda)_3(H_2O)_5]\cdot 3H_2O$ (Ln = Sm (1), Eu (2)). They show a new 2D structural arrangement. Solid-state luminescent properties were also studied.

2. Materials and Methods

All chemicals were reagent grade, purchased from commercial sources, and used without purification. $LnCl_3\cdot 6H_2O$ (Ln = Sm, Eu, 99.9% from Sigma-Aldrich, Burlington, MA, USA) were used as metal sources. Potentiometric measurements were carried out using an automatic titrator Mettler-Toledo DL50-Graphix. Elemental analyses (C, H, and S) were performed on a Thermo FLASH 2000 CHNS/O Analyzer instrument. Infrared spectra were collected as KBr pellets on an FTIR Shimadzu IR-Prestige-21 spectrophotometer from 4000 to 400 cm^{-1}. Thermogravimetric analyses (TGA) were carried out on a Shimadzu TGA-50 instrument with a TA 50I interface, using a platinum cell and nitrogen atmosphere; the experimental conditions were 0.5 °C min^{-1} temperature ramp rate and 50 mL min^{-1} nitrogen flow rate (pure nitrogen was used, water content was less than 3 ppm). Luminescence spectra were recorded from solid crystalline samples using a SHIMADZU RF-5301Pc spectrofluorometer.

2.1. Equilibrium Studies

The standard HCl and NaOH solutions were prepared by diluting Merck standard ampoules. Acid and base stock solutions were standardized against sodium carbonate and potassium hydrogen phthalate, respectively. All solutions were prepared with analytical-grade water (18 µS cm^{-1}) and were freed of carbon dioxide by bubbling with argon. $NaClO_4\cdot H_2O$ (Sigma-Aldrich 98%) was used to adjust the ionic strength of all solutions to 0.15 mol·L^{-1}. The temperature was kept at 25.0 (±0.1) °C. The protonation constants of tda^{2-} were determined by two potentiometric titrations (ca. 150 experimental points each) in the interval 1–8 mmol·L^{-1}. The behavior of the ligand in the presence of either Sm(III) or Eu(III) was then analyzed through three potentiometric titrations (ca. 100–150 experimental points each) at ligand to Ln(III) total molar ratios varying from 1:1 to 3:1. The pH interval from 2.0 to the precipitation of solid Ln(OH)$_3$ in the alkaline region was covered.

In a typical experiment, after thermal equilibrium was reached, hydrogen ion concentrations were determined by successive readings, each performed after an incremental addition of standard 0.1 mol·L^{-1} NaOH solution. Equilibrium attainment after each titrant addition was verified by controlling the deviation of successive e.m.f. readings. Independent stock solutions were used to check reproducibility. The cell electrode potential $E°$

and the acidic junction potential were determined [64] from independent titrations of the strong acid with the titrant solution. The calibration in the alkaline range was checked by recalculating K_w values for each system. The obtained values (average $\log_{10} K_w = 13.7$) were checked to be in line with previously reported data under the same experimental conditions [65]. The formation constant of soluble hydroxo species of Ln(III) was taken from a previous report [66] and was included in the input for the calculation of the formation constants. Further details on data analysis can be found elsewhere [67].

2.2. X-ray Data Collection and Structure Refinement

X-ray diffraction data on single crystals **1** and **2** were collected with an Agilent SuperNOVA diffractometer with microfocus X-ray using Mo Kα radiation (λ = 0.71073 Å). CrysAlisPro [68] software was used to collect, index, scale and apply a numerical absorption correction based on Gaussian integration over a multifaceted crystal model. The structures were solved using ShelXT [69] program using dual methods and refined by full-matrix least-squares minimization on F^2 using ShelXL [70] software. All non-hydrogen atoms were refined anisotropically. Hydrogen atom positions were calculated geometrically and refined using the riding model. The geometrical analysis of the interactions in the structures was performed with PLATON [71] and Olex2 [72] programs. Crystal data, collection procedures, and refinement results are summarized in Table 1.

Table 1. Crystallographic data and structure refinements for compounds **1** and **2**.

Compound	1	2
Formula	$C_{12}H_{28}O_{20}S_3Sm_2$	$C_{12}H_{28}O_{20}S_3Eu_2$
$D_{calc.}/g\,cm^{-3}$	2.231	2.250
μ/mm^{-1}	4.715	5.042
Formula Weight	889.22	892.44
Colour	colorless	colorless
Shape	block-shaped	irregular-shaped
Size/mm³	0.17 × 0.09 × 0.07	0.07 × 0.06 × 0.04
T/K	293(2)	293(2)
Crystal System	triclinic	triclinic
Space Group	P-1	P-1
a/Å	9.0767(3)	9.0706(3)
b/Å	12.1931(4)	12.1653(3)
c/Å	13.3940(4)	13.3578(5)
α/°	63.274(3)	63.364(3)
β/°	88.730(2)	88.684(3)
γ/°	88.545(2)	88.508(3)
V/Å³	1323.47(8)	1317.01(8)
Z	2	2
Z'	1	1
Wavelength/Å	0.71073	0.71073
Radiation type	Mo K_a	Mo K_a
$\theta_{min}/°$	1.702	1.706
$\theta_{max}/°$	28.282	32.043
Measured Refl's.	13,613	17,246
Indep't Refl's	6556	8441
Refl's I ≥ 2 σ(I)	5914	6900
R_{int}	0.0189	0.0276
Parameters	414	407
Restraints	0	0
Largest Peak	0.687	0.951
Deepest Hole	−0.822	−0.920
GooF	1.055	1.035
R_1 (all data) [a]	0.0267	0.0449
R_1 [a]	0.0227	0.0323
wR_2 (all data) [b]	0.0538	0.0653
wR_2 [b]	0.0518	0.0605

[a] $R_1 = \Sigma||F_0|-|F_c||/\Sigma|F_c|$, [b] $wR_2 = \{\Sigma[w(F_0^2 - F_c^2)2]/\Sigma[w(F_0^2)^2]\}^{1/2}$.

Crystallographic data for the structures reported in this contribution have been deposited with the Cambridge Crystallographic Data Centre as supplementary publication 2224921-224922. Copies of the data can be obtained free of charge on application to the CCDC, Cambridge, U.K. (http://www.ccdc.cam.ac.uk/).

2.3. Synthesis of [$Ln_2(tda)_3(H_2O)_5$]·$3H_2O$ (Ln = Sm (1), Eu (2))

1.35 mmol (0.203 g) of H_2tda and 0.45 mmol of $LnCl_3·6H_2O$ (0.164 (1), 0.165 (2) g) were dissolved in 10 mL of water at room temperature with continuous stirring. Then, the pH of the solution was adjusted to ca. 3.3 with aqueous 0.5 M NaOH. If a small amount of a white solid remained at this point, it was filtered through paper and discarded. The clear solution was allowed to evaporate slowly. After 10 days, a crystalline material was formed, filtrated, washed with two portions of 1 mL of water, and air-dried. Some crystals were suitable for single-crystal X-ray diffraction analysis. Yield: 52% (1), 61% (2). Elemental analysis (%) Calcd. for 1, $C_{12}H_{28}Sm_2O_{20}S_3$: C, 16.21; H, 3.17; S, 10.82. Found: C, 16.55; H, 3.00; S, 11.09. Calcd. for 2, $C_{12}H_{28}Eu_2O_{20}S_3$: C, 16.15; H, 3.16; S, 10.78. Found: C, 16.42, H, 2.81, S, 11.03. Found: C, 16.42, H, 2.81, S, 11.45. IR (KBr, cm^{-1}): main signals are almost identical for compounds 1 and 2: 3598(m), 3518(m), 3362(s), 2984(m), 2918(m), 1595(sh), 1564(vs), 1426(s), 1383(s), 1229(s), 1217(w), 1157(m), 1130(m), 962(w), 952(m), 918(m), 899(m), 733(m), 721(m), 710(m), 605(s), 463(m).

3. Results

3.1. Solution Studies

For a rational design of the synthetic procedure, the first step of the study was to look at the Ln(III)-tda systems in solution at room temperature and low ionic strength since previously reported data were not conclusive about the species formed, especially in the acid interval in which protonated species were detected by some authors but not by others (Table S1). Protonation equilibrium constants of the ligand (Table 2) were also redetermined under identical conditions: 0.15 mol·L^{-1} $NaClO_4$ at 25.0 °C. The obtained results for the protonation constants are in total agreement with previous reports under similar conditions [65,73,74]. With these results and the previously reported hydrolysis constants of the Ln(III) ions under similar conditions [66], the stability constants of the species Ln(III)-tda were determined. This is also shown in Table 2. Only cationic species 1:1 ([Ln(tda)]$^+$ and [Ln(Htda)]$^{2+}$) and the anionic species 1:2 ($Ln(tda)_2]^-$) were detected in solution, with similar stability constant values for Sm(III) and Eu(III). It is interesting to compare the stability constants of tda species with those for oda and ida (some of them are also included in Table 2). A close inspection of $\log_{10} K$ values shows that Ln(III)-tda species are much less stable than the analogous complexes with oda or ida. The change of O or N by S as a donor atom represents a loss of stability in the coordination compounds of lanthanide(III) ions.

Table 2. Logarithm of the acid-base and complexation equilibrium constants determined in this work in 0.15 mol·L^{-1} $NaClO_4$ at 25.0 °C. H_2tda represents the fully protonated neutral form of thiodiacetic acid. Values given in parentheses are the 1σ statistical uncertainties in the last digit of the determined constant values. Selected stability constants for analogous oda and ida complexes (again, H_2oda and H_2ida represent the fully protonated neutral forms) were taken from selected reported data under similar experimental conditions.

Equilibrium	$\log_{10} K$	σ
$tda^{2-} + H^+ \rightarrow Htda^-$	4.23(1)	1.3
$tda^{2-} + 2H^+ \rightarrow H_2tda$	7.28(2)	
$Sm^{3+} + tda^{2-} \rightarrow [Sm(tda)]^+$	2.94(4)	0.7
$Sm^{3+} + 2tda^{2-} \rightarrow [Sm(tda)_2]^-$	5.16(6)	
$Sm^{3+} + H^+ + tda^{2-} \rightarrow [Sm(Htda)]^{2+}$	6.28(5)	

Table 2. Cont.

Equilibrium	$\log_{10} K$	σ
$Eu^{3+} + tda^{2-} \rightarrow [Eu(tda)]^+$	3.06(1)	
$Eu^{3+} + 2tda^{2-} \rightarrow [Eu(tda)_2]^-$	5.96(5)	0.4
$Eu^{3+} + H^+ + tda^{2-} \rightarrow [Eu(Htda)]^{2+}$	5.7(2)	
oda		Ref.
$Sm^{3+} + oda^{2-} \rightarrow [Sm(oda)]^+$	5.64	[75]
$Sm^{3+} + 2oda^{2-} \rightarrow [Sm(oda)_2]^-$	9.62	[75]
$Eu^{3+} + oda^{2-} \rightarrow [Eu(oda)]^+$	5.53	[76]
$Eu^{3+} + 2oda^{2-} \rightarrow [Eu(oda)_2]^-$	10.04	[76]
ida		
$Sm^{3+} + ida^{2-} \rightarrow [Sm(ida)]^+$	5.914	[14]
$Sm^{3+} + 2ida^{2-} \rightarrow [Sm(ida)_2]^-$	10.230	[14]
$Eu^{3+} + ida^{2-} \rightarrow [Eu(ida)]^+$	6.48	[77]
$Eu^{3+} + 2ida^{2-} \rightarrow [Eu(ida)_2]^-$	11.65	[77]

Figure 1 shows the species distribution diagrams for Sm(III)-tda system built with these results for the ligand-to-metal molar ratios 1:1 and 3:1, while Figure S1 shows similar diagrams for Eu(III). The low stability of these complexes is reflected in the high percentage of free Ln(III), especially below pH 4–5. Even though the complex species $[Ln(Htda)]^{2+}$ is detected in this interval (contrary to what happens in oda or ida-containing systems), the partially protonated ligand gives place to a low-stability species that forms only in a relatively low percentage (the calculated $\log_{10} K$ values for $Ln^{3+} + Htda^- \rightarrow [Ln(Htda)]^{2+}$ are 2.0 and 1.5 for Sm and Eu, respectively). It is worth mentioning at this point that the present results are in perfect agreement with previous findings based on the luminescence lifetime measurements of Eu(III) or Sm(III) ions in the presence of tda in an aqueous solution [78,79]. In particular, these findings suggest the non-participation of the S atom in the coordination sphere, which is in agreement with the formation of much less stable species in comparison to what happens with the analogous oda or ida ligands [78,79]. Besides, similar experiments carried out in the acid interval show the same average number of ca. six Sm-coordinated water molecules for conditions in which either $[Sm(tda)]^+$, $[Sm(Htda)]^{2+}$ or $[Sm(tda)(Htda)]$ should predominate [78,80]. This accounts for the non-existence of the last species, which was not detected in this work. On the other hand, the hydrolysis of the lanthanide(III) ion at pH values above 7–8 represents a relevant competitive process to be considered in the synthesis of the compounds. In that sense, just above pH 3 and in the presence of ligand excess, mononuclear species are expected to be present. In contrast, the competition of hydrolysis processes is expected to be minimized.

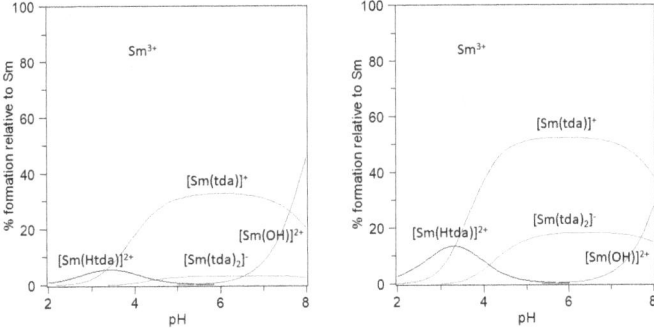

Figure 1. Species distribution diagram of the Sm-tda system at 25.0 °C, I = 0.15 mol·L^{-1} NaClO$_4$. Left: total $[Sm^{3+}]$ = 1 mM and total [tda] = 1 mM. Right: total $[Sm^{3+}]$ = 1 mM and total [tda] = 3 mM.

3.2. Synthesis and Characterization

Following our previous strategy to isolate Ln complexes with ida and derivatives [19,81] and taking into account the results of the preceding section, we prepared the complexes [Ln$_2$(tda)$_3$(H$_2$O)$_5$]·3H$_2$O (Ln = Sm (**1**), Eu (**2**)). An aqueous solution of LnCl$_3$ and H$_2$tda (molar ratio 1:3, pH 3.3) was allowed to evaporate slowly to obtain the crystalline compounds. Preliminary characterization of the solids by elemental analysis is in good agreement with the proposed formula. The IR spectra (Figure S2) were almost identical for both complexes suggesting very similar structures. It is noticeable the shift and splitting of the sharp signals ν_s(COO) (1698 cm^{-1}) and ν_{as}(COO) (1430 cm^{-1}) of the free ligand: **1** and **2** exhibit very strong ν_s(COO) signals at 1595 and 1564 cm^{-1} and ν_{as}(COO) at 1425 and 1383 cm^{-1}. The stoichiometric ratio found in the solids (molar ratio Ln:tda 2:3) is the same as found in the complexes [Ln$_2$(tda)$_3$(H$_2$O)$_2$] previously reported [57–59]. In those previous reports, the synthesis was performed by solvothermal procedures and, in general, in water: ethanol mixtures.

TGA of the solids (Figure S3) shows a broad weight loss up to 200 °C, corresponding to all the water molecules (calculated for **1**, 16.2%, found 15.2%; calculated for **2**, 16.2%, found 15.3%). Decomposition appears in both complexes at ca. 320 °C.

3.3. Crystal Structures

It was possible to obtain single crystals of **1** and **2**, which crystallize in the triclinic space group P$\bar{1}$. They are isostructural, so we will only discuss the structure of **1**. Selected bond lengths are presented in Table 3. Two crystallographically non-equivalent Sm atoms are present, both with coordination number nine (Figure 2). Sm1 atom is surrounded by seven carboxylic O atoms, one S atom (S1), and one O of a coordinated water molecule. Sm2 is bound to four O atoms from water molecules and five carboxylic O atoms arising from two different ligands (those containing the non-coordinated S2 and S3 atoms). Sm1-S1 distance is 3.130(1) Å, which is close to the values found in similar structures containing Ln(III) ions and thiol-type S atoms (Table S2).

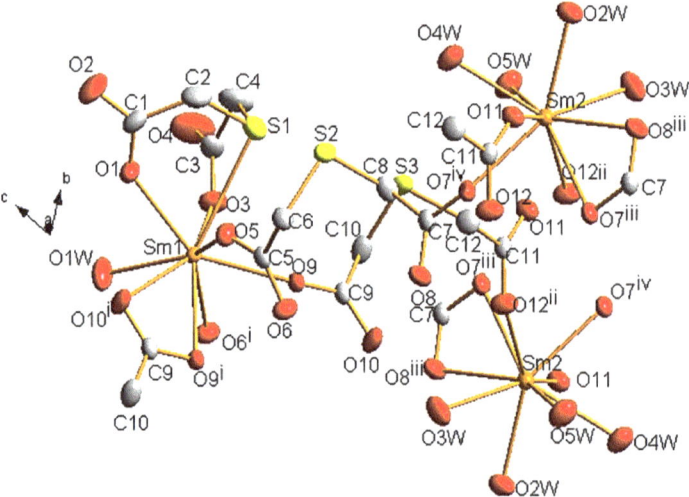

Figure 2. Perspective drawing of **1** showing the atom labels. Thermal ellipsoids are plotted at the 50% probability level. Hydrogen atoms and crystallization water molecules are omitted for clarity. Color code: Sm, orange; C, light grey; O, red; S, yellow. The symmetry-related atoms were obtained by applying the symmetry codes [i] 1 − x, −y, 1 − z; [ii] −x, 1 − y, −z; [iii] 1 − x, 1 − y, −z; [iv] −1 + x, +y, +z.

Table 3. Bond lengths (Å) around central atom for **1** and **2**.

	1				2		
Sm1-O9	2.392(2)	Sm2-O7 [iii]	2.596(2)	Eu1-O10 [i]	2.414(3)	Eu2-O11 [iv]	2.580(2)
Sm1-O9 [i]	2.585(2)	Sm2-O7 [iv]	2.410(2)	Eu1-O6 [i]	2.378(2)	Eu2-O8 [ii]	2.392(3)
Sm1-O6 [i]	2.428(2)	Sm2-O8 [iii]	2.552(2)	Eu1-O9	2.387(3)	Eu2-O11	2.394(2)
Sm1-O10 [i]	2.545(3)	Sm2-O11	2.400(2)	Eu1-O5	2.539(3)	Eu2-O7 [iii]	2.405(3)
Sm1-O1	2.376(2)	Sm2-O12 [ii]	2.413(2)	Eu1-O1	2.362(3)	Eu2-O12 [ii]	2.544(3)
Sm1-O5	2.396(2)	Sm2-O2W	2.433(3)	Eu1-O3	2.361(3)	Eu2-O5W	2.420(3)
Sm1-O3	2.376(2)	Sm2-O3W	2.438(2)	Eu1-O6	2.570(3)	Eu2-O2W	2.444(3)
Sm1-O1W	2.427(3)	Sm2-O5W	2.456(2)	Eu1-O1W	2.423(3)	Eu2-O4W	2.503(3)
Sm1-S1	3.130(1)	Sm2-O4W	2.520(3)	Eu1-S1	3.126(1)	Eu2-O3W	2.427(3)

[i] $1-x, -y, 1-z$; [ii] $-x, 1-y, -z$; [iii] $1-x, 1-y, -z$; [iv] $-1+x, +y, +z$
[i] $1-x, 2-y, 1-z$; [ii] $-x, 2-y, 1-z$; [iii] $x, -1+y, 1+z$; [iv] $-x, 1-y, 2-z$

In both Sm ions, polyhedra can be described as a tricapped trigonal prism (JTCTPR-9) [82,83]. For Sm1, the tricapped trigonal prism is quite irregular with O1, O3, O5, O9, O6i, and O10i as the vertices of the prism (average distance 2.419 Å), and OW1, O9i, and S1 in the apices (average distance 2.714 Å). In the case of Sm2, O2W, O3W, O4W, O11, O7iii, and O12ii conform the prism (average distance 2.467 Å), while O5W, O7iv, and O8iii occupy the apical positions (average distance 2.473 Å). This is shown in Figure S4.

It is interesting to view the connection between Sm(III) ions provided by the ligand (Scheme 1). Three non-equivalent ligands are present in the structure. Tda residue containing S1 does not link Sm ions. On the contrary, it is only bound to Sm1 as a tridentate ligand, generating two O atoms (O2 and O6) that do not participate in the coordination. The ligand with S2 connects four Sm ions in a monodentate fashion. The third ligand (with S3) connects three Sm(III) ions, bis-monodentate towards Sm1 and bidentate towards Sm2.

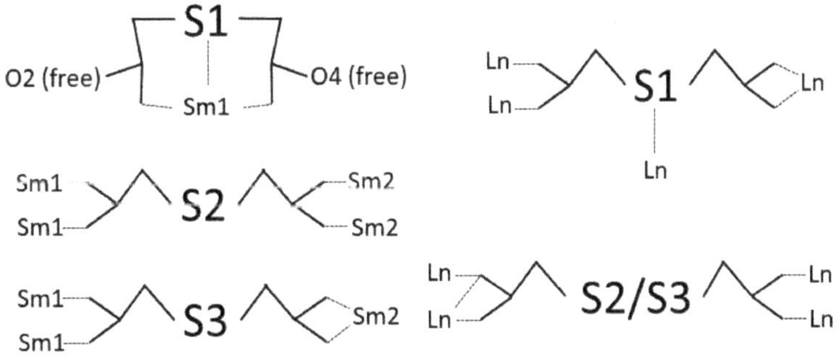

Scheme 1. Coordination modes of tda in compound **1** (**left**) and in previously reported structures [Ln$_2$(tda)$_3$(H$_2$O)$_2$] (**right**) [59–61].

The coordinative versatility of this ligand provokes the formation of thick sheets in the bc plane (Figure 3). Sm1 polyhedra are disposed of in couples sharing an edge through carboxylate groups of ligands with S2 and S3. In contrast, the Sm2 polyhedron shares an edge with an Sm1 polyhedron through O atoms from the third ligand (with S3).

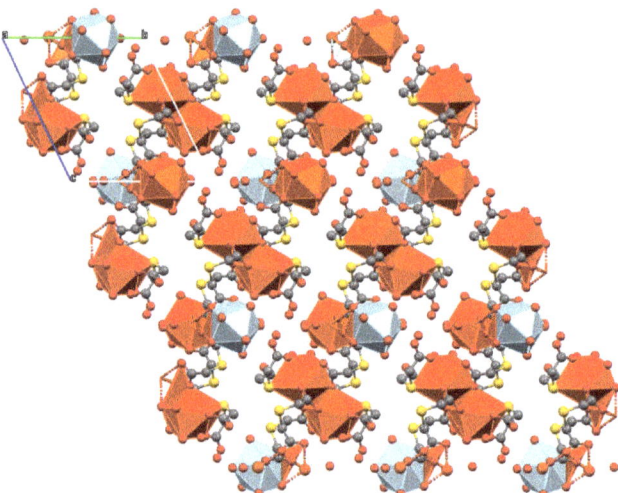

Figure 3. Packing of **1** in the bc plane. Polyhedra of Sm1 are colored orange, while Sm2 ones are light blue. Hydrogen atoms and crystallization water molecules are omitted for clarity.

Crystallization water molecules occupy the free space between the sheets and are involved in H-bonds, in particular with the uncoordinated O atoms. This is shown in Figure S5.

Previously reported structures with formula $[Ln_2(tda)_3(H_2O)_2]$ exhibit the identical molar ratio Ln:tda found in **1** and **2** [59–61]. They crystallize in the same triclinic space group $P\bar{1}$, contain two crystallographically non-equivalent Ln atoms, and also present a 2D structure. It is interesting to compare the Ln-S distances. They are 3.105 and 3.091 Å for Dy [59,60], 3.105 Å for Eu [61], 3.098 Å for Gd [60], 3.122 Å for Sm [59], and 3.099 Å for Tb [60]. These values have not been considered as a bond distance, except in the Eu structure. Assuming that an Ln-S bond is present, Ln1 is coordinated by eight oxygen atoms (all of them from carboxylate groups) and one sulfur atom, and Ln2 is coordinated by nine O atoms, two of them from coordinated water molecules. From the ligand point of view, two coordination modes are present, as shown in Scheme 1.

3.4. Photophysical Studies

Figure 4 depicts the solid-state luminescence spectra of compound **2** upon direct excitation of Eu(III) ion at 394 nm, selected from the excitation spectrum and giving place to the characteristic emission profile of the intra-configuration emission of the metal ion. Eu(III) is, in general, a much more intense luminescence emitter than Sm(III) [84], accounting for compound **2** is a much more intense emitter than compound **1** (which is shown in Figure S6, together with the excitation spectra and the assignment for emission bands). The emission spectra of Eu(III) show the characteristic five main bands corresponding to the intra-configuration transitions $^5D_0 \rightarrow {}^7F_J$, with J = 0–4. Noticeably, the truly forbidden transition $^5D_0 \rightarrow {}^7F_0$ is observed, even though with very low intensity. This band is probably associated with the less-symmetric Eu1 center (bound S1 atom) described in the crystal structure. The presence of the S1 atom in one of the apices of the tricapped trigonal prism excludes the existence of a mirror plane orthogonal to the main symmetry axis [85,86]. Indeed, in the $[Eu(oda)_3]^{3-}$ complex, with all positions of the tricapped trigonal prism occupied O atoms and similar Eu-O bond distances, this forbidden band is not observed [18]. The magnetic dipole transition $^5D_0 \rightarrow {}^7F_1$ band shows in compound **2** an intense degenerated profile, in line with the rotational symmetry in the coordination geometry. Furthermore, the $^5D_0 \rightarrow {}^7F_2$ hypersensitive band is 1.3 times more intense than the magnetic dipole transition $^5D_0 \rightarrow {}^7F_1$ band. Noticeably, in the previously reported complex [Eu$_2$(tda)$_3$(H$_2$O)$_2$], with just two coordinated water molecules coordinated to one of the emissive centers, the $^5D_0 \rightarrow {}^7F_2$ hypersensitive band shows an intensity more than three times higher than that of the $^5D_0 \rightarrow {}^7F_1$ band [61]. The presence of four coordinated water molecules in **2** accounts for the lower comparative observed intensity of the $^5D_0 \rightarrow {}^7F_2$ band. Also in line with this, the behavior of compound **1** is also influenced by the presence of coordinated water molecules, which diminishes the emissive behavior of Sm ion, relative to that observed for [Sm$_2$(tda)$_3$(H$_2$O)$_2$] [59] (Figure S6).

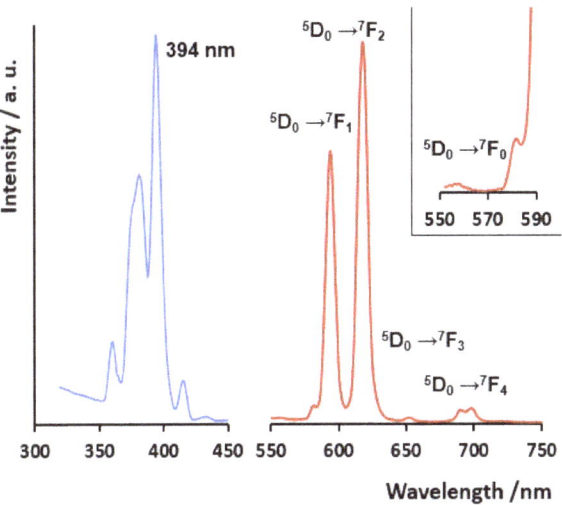

Figure 4. (**Left**): solid-state excitation spectra. (**Right**): emission spectra of compound **2** excited at 394 nm. Transitions assigned to each band are also shown. An inset is included for more detail on the $^5D_0 \rightarrow {}^7F_0$ band.

4. Conclusions

The chemistry of thiodiacetato with the lanthanide ions Sm(III) and Eu(III) has been explored. First, the solution behavior of this system has been studied, and some light has been shed on the formed species, considering also previously reported inconclusive findings. Starting from this knowledge, two new compounds have been obtained and fully characterized, showing the versatility of the thiodiacetato ligand, which can either chelate or connect the lanthanide(III) centers. The solid-state optical properties of the two compounds were studied, the Eu(III) compound resulting in a better luminescence emitter material.

Supplementary Materials: The following supporting information can be downloaded at: https://www.mdpi.com/article/10.3390/cryst13010056/s1, Figure S1: Species distribution diagram of the Eu-tda system at 25.0 °C, I = 0.15 mol·L^{-1} NaClO$_4$; Figure S2: IR spectra of complexes **1** and **2**, together with the protonated ligand; Figure S3: TGA diagram of compounds **1** and **2** under nitrogen atmosphere; Figure S4: Polyhedra around the Sm(III) ions in **1**; Figure S5: Packing of **1** in the *bc* plane showing H-bonds; Figure S6: solid-state excitation and emission spectra of compound **1**; Table S1: Previously potentiometrically determined stability constant values for Sm and Eu complexes with tda; Table S2: Typical bond distances Ln-S found in structures containing Ln(III) ions and thiol ligands.

Author Contributions: Conceptualization, C.K.; synthesis and characterization, C.K.; potentiometric studies, J.T.; photophysical studies, J.T.; diffraction studies, J.G.-P.; writing—original draft preparation, C.K. and J.T.; writing—review and editing, C.K., J.T and J.G.-P. All authors have read and agreed to the published version of the manuscript.

Funding: This research received no external funding.

Data Availability Statement: Not applicable.

Acknowledgments: We are grateful for the financial support from the Uruguayan organizations CSIC (Comisión Sectorial de Investigación Científica) by Programa de Apoyo a Grupos de Investigación and PEDECIBA (Programa para el Desarrollo de las Ciencias Básicas). This work has also been partially supported by MCIN/AEI/10.13039/5011000011033 in the project PID2019-106383GB-C44. J.G.-P. thanks to Servicios Generales de Apoyo a la Investigación (SEGAI) at La Laguna University.

Conflicts of Interest: The authors declare no conflict of interest.

References

1. Murugesu, M.; Schelter, E.J. Not Just Lewis Acids: Preface for the Forum on New Trends and Applications for Lanthanides. *Inorg. Chem.* **2016**, *55*, 9951–9953. [CrossRef] [PubMed]
2. Cotton, S.A.; Raithby, P.R. Systematics and Surprises in Lanthanide Coordination Chemistry. *Coord. Chem. Rev.* **2017**, *340*, 220–231. [CrossRef]
3. Chundawat, N.S.; Jadoun, S.; Zarrintaj, P.; Chauhan, N.P.S. Lanthanide Complexes as Anticancer Agents: A Review. *Polyhedron* **2021**, *207*, 115387. [CrossRef]
4. Bao, G. Lanthanide Complexes for Drug Delivery and Therapeutics. *J. Lumin.* **2020**, *228*, 117622. [CrossRef]
5. Zheng, Z.; Lu, H.; Wang, Y.; Bao, H.; Li, Z.-J.; Xiao, G.-P.; Lin, J.; Qian, Y.; Wang, J.-Q. Tuning of the Network Dimensionality and Photoluminescent Properties in Homo- and Heteroleptic Lanthanide Coordination Polymers. *Inorg. Chem.* **2021**, *60*, 1359–1366. [CrossRef]
6. Zhang, P.; Guo, Y.-N.; Tang, J. Recent Advances in Dysprosium-Based Single Molecule Magnets: Structural Overview and Synthetic Strategies. *Coord. Chem. Rev.* **2013**, *257*, 1728–1763. [CrossRef]
7. Jin, J.; Xue, J.; Liu, Y.; Yang, G.; Wang, Y.-Y. Recent Progresses in Luminescent Metal–Organic Frameworks (LMOFs) as Sensors for the Detection of Anions and Cations in Aqueous Solution. *Dalton Trans.* **2021**, *50*, 1950–1972. [CrossRef]
8. Huangfu, M.; Wang, M.; Lin, C.; Wang, J.; Wu, P. Luminescent Metal–Organic Frameworks as Chemical Sensors Based on "Mechanism–Response": A Review. *Dalton Trans.* **2021**, *50*, 3429–3449. [CrossRef]
9. Kremer, C.; Torres, J.; Domínguez, S. Lanthanide Complexes with Oda, Ida, and Nta: From Discrete Coordination Compounds to Supramolecular Assemblies. *J. Mol. Struct.* **2008**, *879*, 130–149. [CrossRef]
10. Wen, Y.-H.; Wu, X.-H.; Bi, S.; Zhang, S.-S. Synthesis and Structural Characterization of Lanthanide Oxalate–Oxydiacetate Mixed-Ligand Coordination Polymers {[Ln(Oda)(H_2O)$_x$]$_2$(Ox)}$_n$ (x = 3 for Ln = La, Ce, Pr, Gd, Tb and x = 2 for Ln = Er). *J. Coord. Chem.* **2009**, *62*, 1249–1259. [CrossRef]
11. Lennartson, A.; Håkansson, M. Total Spontaneous Resolution of Nine-Coordinate Complexes. *CrystEngComm* **2009**, *11*, 1979. [CrossRef]
12. Zhou, Q.; Yang, F.; Liu, D.; Peng, Y.; Li, G.; Shi, Z.; Feng, S. Synthesis, Structures, and Magnetic Properties of Three Fluoride-Bridged Lanthanide Compounds: Effect of Bridging Fluoride Ions on Magnetic Behaviors. *Inorg. Chem.* **2012**, *51*, 7529–7536. [CrossRef] [PubMed]
13. Ma, J.; Ma, T.; Qian, R.; Zhou, L.; Guo, Q.; Yang, J.-H.; Yang, Q. Na–Ln Heterometallic Coordination Polymers: Structure Modulation by Na^+ Concentration and Efficient Detection to Tetracycline Antibiotics and 4-(Phenylazo)Aniline. *Inorg. Chem.* **2021**, *60*, 7937–7951. [CrossRef]
14. Torres, J.; Kremer, C.; Domínguez, S. Chemical Speciation of Polynuclear Complexes Containing [$Ln_2M_3L_6$] Units. *Pure Appl. Chem.* **2008**, *80*, 1303–1316. [CrossRef]
15. Cancino, P.; Santibañez, L.; Stevens, C.; Fuentealba, P.; Audebrand, N.; Aravena, D.; Torres, J.; Martinez, S.; Kremer, C.; Spodine, E. Influence of the Channel Size of Isostructural 3d–4f MOFs on the Catalytic Aerobic Oxidation of Cycloalkenes. *New J. Chem.* **2019**, *43*, 11057–11064. [CrossRef]
16. Santibáñez, L.; Escalona, N.; Torres, J.; Kremer, C.; Cancino, P.; Spodine, E. CuII- and CoII-Based MOFs: {[La_2Cu_3(μ-H_2O)(ODA)$_6$(H_2O)$_3$]·3H_2O}$_n$ and {[La_2Co_3(ODA)$_6$(H_2O)$_6$]·12H_2O}$_n$. The Relevance of Physicochemical Properties on the Catalytic Aerobic Oxidation of Cyclohexene. *Catalysts* **2020**, *10*, 589. [CrossRef]
17. Igoa, F.; Peinado, G.; Suescun, L.; Kremer, C.; Torres, J. Design of a White-Light Emitting Material Based on a Mixed-Lanthanide Metal Organic Framework. *J. Solid State Chem.* **2019**, *279*, 120925. [CrossRef]
18. Igoa, F.; Romero, M.; Peinado, G.; Castiglioni, J.; Gonzalez-Platas, J.; Faccio, R.; Suescun, L.; Kremer, C.; Torres, J. Ln(III)–Ni(II) Heteropolynuclear Metal Organic Frameworks of Oxydiacetate with Promising Proton-Conductive Properties. *CrystEngComm* **2020**, *22*, 5638–5648. [CrossRef]
19. Kremer, C.; Morales, P.; Torres, J.; Castiglioni, J.; González-Platas, J.; Hummert, M.; Schumann, H.; Domínguez, S. Novel Lanthanide–Iminodiacetate Frameworks with Hexagonal Pores. *Inorg. Chem. Commun.* **2008**, *11*, 862–864. [CrossRef]
20. Bonomo, R.P.; Rizzarelli, E.; Bresciani-Pahor, N.; Nardin, G. Properties and X-Ray Crystal Structures of Copper(II) Mixed Complexes with Thiodiacetate and 2,2′-Bipyridyl or 2,2′:6′2″-Terpyridyl. *J. Chem. Soc. Dalton Trans.* **1982**, 681–685. [CrossRef]
21. Alarcón-Payer, C.; Pivetta, T.; Choquesillo-Lazarte, D.; González-Pérez, J.M.; Crisponi, G.; Castiñeiras, A.; Niclós-Gutiérrez, J. Thiodiacetato-Copper(II) Chelates with or without N-Heterocyclic Donor Ligands: Molecular and/or Crystal Structures of [Cu(Tda)]n, [Cu(Tda)(Him)$_2$(H_2O)] and [Cu(Tda)(5Mphen)]·2H_2O (Him=imidazole, 5Mphen=5-Methyl-1,10-Phenanthroline). *Inorg. Chim. Acta* **2005**, *358*, 1918–1926. [CrossRef]

22. Abbaszadeh, A.; Safari, N.; Amani, V.; Notash, B.; Raei, F.; Eftekhar, F. Mononuclear and Dinuclear Copper(II) Complexes Containing N, O and S Donor Ligands: Synthesis, Characterization, Crystal Structure Determination and Antimicrobial Activity of [Cu(Phen)(Tda)]·2H$_2$O and [(Phen)$_2$Cu(μ-Tda)Cu(Phen)](ClO$_4$)$_2$·1.5H$_2$O. *Iran. J. Chem. Chem. Eng. IJCCE* **2014**, *33*, 1–33. [CrossRef]
23. Patel, D.K.; Choquesillo-Lazarte, D.; Domínguez-Martín, A.; Brandi-Blanco, M.P.; González-Pérez, J.M.; Castiñeiras, A.; Niclós-Gutiérrez, J. Chelating Ligand Conformation Driving the Hypoxanthine Metal Binding Patterns. *Inorg. Chem.* **2011**, *50*, 10549–10551. [CrossRef] [PubMed]
24. Kopel, P.; Trávníček, Z.; Marek, J.; Mrozinski, J. Syntheses and Study on Nickel(II) Complexes with Thiodiglycolic Acid and Nitrogen-Donor Ligands. X-Ray Structures of [Ni(Bpy)(Tdga)(H$_2$O)]·4H$_2$O and [(En)Ni(μ-Tdga)$_2$Ni(En)]·4H$_2$O (TdgaH$_2$=thiodiglycolic Acid). *Polyhedron* **2004**, *23*, 1573–1578. [CrossRef]
25. Wang, Y.-L.; Chang, G.-J.; Liu, B.-X. Aqua(2,2′-Diamino-4,4′-Bi-1,3-Thiazole-κ^2 N^3,$N^{3'}$)(Thiodiacetato-κ^3 O,S,O')Nickel(II) Monohydrate. *Acta Crystallogr. Sect. E Struct. Rep. Online* **2011**, *67*, m681. [CrossRef]
26. Pan, T.-T.; Su, J.-R.; Xu, D.-J. Hexaaquanickelate(II) Bis(Thiodiacetato-κ^3 O,S,O')Nickel(II) Tetrahydrate. *Acta Crystallogr. Sect. E Struct. Rep. Online* **2005**, *61*, m1376–m1378. [CrossRef]
27. Alarcón-Payer, C.; Pivetta, T.; Choquesillo-Lazarte, D.; González-Pérez, J.M.; Crisponi, G.; Castiñeiras, A.; Niclós-Gutiérrez, J. Structural Correlations in Nickel(II)–Thiodiacetato Complexes: Molecular and Crystal Structures and Properties of [Ni(Tda)(H$_2$O)$_3$]. *Inorg. Chem. Commun.* **2004**, *7*, 1277–1280. [CrossRef]
28. Pan, T.-T.; Su, J.-R.; Xu, D.-J. Tris(1 H -Imidazole-κ N^3)(Thiodiacetato-κ^3 O,S,O')Nickel(II) Monohydrate. *Acta Crystallogr. Sect. E Struct. Rep. Online* **2005**, *61*, m1576–m1578. [CrossRef]
29. Delaunay, J.; Kappenstein, C.; Hugel, R. Structure Cristalline et Moléculaire Du Bis-Thio(Diacétato)Nickelate(II) de Potassium Trihydraté. *Acta Crystallogr. B* **1976**, *32*, 2341–2345. [CrossRef]
30. Cao, L.; Liu, J.-G.; Xu, D.-J. Tris(1 H -Benzimidazole-κ N^3)(Thiodiacetato-κ^3 O,S,O')Cobalt(II) Dihydrate. *Acta Crystallogr. Sect. E Struct. Rep. Online* **2006**, *62*, m579–m581. [CrossRef]
31. Liu, B.-X.; Yu, J.-Y.; Xu, D.-J. Aqua(2,2′-Diamino-4,4′-Bi-1,3-Thiazole-κ^2 N, N')(Thiodiacetato-κ^3 O,S,O')Cobalt(II) Dihydrate. *Acta Crystallogr. Sect. E Struct. Rep. Online* **2005**, *61*, m1978–m1980. [CrossRef]
32. Korchagin, D.V.; Gureev, Y.E.; Yureva, E.A.; Shilov, G.V.; Akimov, A.V.; Misochko, E.Y.; Morgunov, R.B.; Zakharov, K.V.; Vasiliev, A.N.; Palii, A.V.; et al. Field-Induced Single-Ion Magnet Based on a Quasi-Octahedral Co(II) Complex with Mixed Sulfur–Oxygen Coordination Environment. *Dalton Trans.* **2021**, *50*, 13815–13822. [CrossRef] [PubMed]
33. Grirrane, A.; Pastor, A.; Álvarez, E.; Mealli, C.; Ienco, A.; Masi, D.; Galindo, A. Thiodiacetate Cobalt(II) Complexes: Synthesis, Structure and Properties. *Inorg. Chem. Commun.* **2005**, *8*, 463–466. [CrossRef]
34. Grirrane, A.; Pastor, A.; Álvarez, E.; Mealli, C.; Ienco, A.; Rosa, P.; Galindo, A. Thiodiacetate and Oxydiacetate Cobalt Complexes: Synthesis, Structure and Stereochemical Features. *Eur. J. Inorg. Chem.* **2007**, *2007*, 3543–3552. [CrossRef]
35. Wang, H.; Gao, S.; Ng, S.W. Hexaaquacobalt(II) Bis(2,2′-Sulfanediyldiacetato-κ^3 O,S,O')Cobaltate(II) Tetrahydrate. *Acta Crystallogr. Sect. E Struct. Rep. Online* **2011**, *67*, m1521. [CrossRef] [PubMed]
36. Wu, J.-Y.; Xie, L.-M.; He, H.-Y.; Zhou, X.; Zhu, L.-G. Aqua(1,10-Phenanthroline)(Thiodiglycolato)Cobalt(II). *Acta Crystallogr. Sect. E Struct. Rep. Online* **2005**, *61*, m568–m570. [CrossRef]
37. Grirrane, A.; Pastor, A.; Galindo, A.; Álvarez, E.; Mealli, C.; Ienco, A.; Orlandini, A.; Rosa, P.; Caneschi, A.; Barra, A.; et al. Thiodiacetate–Manganese Chemistry with N Ligands: Unique Control of the Supramolecular Arrangement over the Metal Coordination Mode. *Chem. Eur. J.* **2011**, *17*, 10600–10617. [CrossRef]
38. Drew, M.G.B.; Rice, D.A.; Timewell, C.W. Crystal and Molecular Structure of Triaquazinc(II) Thiodiglycolate Monohydrate. *J. Chem. Soc. Dalton Trans.* **1975**, 144–148. [CrossRef]
39. Baggio, R.; Perec, M.; Garland, M.T. Aqua(2,2′-Bipyridyl-N,N′)(Thiodiacetato-O,O′,S)Zinc(II) Tetrahydrate. *Acta Crystallogr. C* **1996**, *52*, 2457–2460. [CrossRef]
40. Arıcı, M.; Yeşilel, O.Z.; Acar, E.; Dege, N. Synthesis, Characterization and Properties of Nicotinamide and Isonicotinamide Complexes with Diverse Dicarboxylic Acids. *Polyhedron* **2017**, *127*, 293–301. [CrossRef]
41. Pan, T.-T.; Xu, D.-J. Tris(1H-Benzimidazole-κ N^3)(Thiodiacetato-κ^3O,S,O')Cadmium(II) Dihydrate. *Acta Crystallogr. Sect. E Struct. Rep. Online* **2005**, *61*, m1735–m1737. [CrossRef]
42. Grirrane, A.; Pastor, A.; Álvarez, E.; Galindo, A. Magnesium Dicarboxylates: First Structurally Characterized Oxydiacetate and Thiodiacetate Magnesium Complexes. *Inorg. Chem. Commun.* **2005**, *8*, 453–456. [CrossRef]
43. Shan, X.; Ellern, A.; Guzei, I.A.; Espenson, J.H. Syntheses and Oxidation of Methyloxorhenium(V) Complexes with Tridentate Ligands. *Inorg. Chem.* **2003**, *42*, 2362–2367. [CrossRef] [PubMed]
44. Álvarez, L.; Grirrane, A.; Moyano, R.; Álvarez, E.; Pastor, A.; Galindo, A. Comparison of the Coordination Capabilities of Thiodiacetate and Oxydiacetate Ligands through the X-Ray Characterization and DFT Studies of [V(O)(Tda)(Phen)]·4H$_2$O and [V(O)(Oda)(Phen)]·1.5H$_2$O. *Polyhedron* **2010**, *29*, 3028–3035. [CrossRef]
45. Zangl, A.; Klüfers, P.; Schaniel, D.; Woike, T. Photoinduced Linkage Isomerism of {RuNO}6 Complexes with Bioligands and Related Chelators. *Dalton Trans.* **2009**, 1034–1045. [CrossRef] [PubMed]
46. Marek, J.; Trávníček, Z.; Kopel, P. Diaquabis(1,10-Phenanthroline-κ^2 N,N')Manganese(II) Thiodiglycolate Bis(1,10-Phenanthroline-κ^2 N,N')(Thiodiglycolato-κ^2 O,O')Manganese(II) Tridecahydrate. *Acta Crystallogr. C* **2003**, *59*, m429–m431. [CrossRef]

47. Wang, L.; Shan, Y.; Gu, X.; Ni, L.; Zhang, W. Assembly and Photocatalysis of Three Novel Metal–Organic Frameworks Tuned by Metal Polymeric Motifs. *J. Coord. Chem.* **2015**, *68*, 2014–2028. [CrossRef]
48. Baggio, R.; Garland, M.T.; Manzur, J.; Peña, O.; Perec, M.; Spodine, E.; Vega, A. A Dinuclear Copper(II) Complex Involving Monoatomic O-Carboxylate Bridging and Cu–S(Thioether) Bonds: [Cu(Tda)(Phen)]$_2$·H$_2$tda (Tda=thiodiacetate, Phen=phenanthroline). *Inorg. Chim. Acta* **1999**, *286*, 74–79. [CrossRef]
49. Ahmad, M.S.; Khalid, M.; Khan, M.S.; Shahid, M.; Ahmad, M.; Monika; Ansari, A.; Ashafaq, M. Exploring Catecholase Activity in Dinuclear Mn(II) and Cu(II) Complexes: An Experimental and Theoretical Approach. *New J. Chem.* **2020**, *44*, 7998–8009. [CrossRef]
50. Kopel, P.; Trávníček, Z.; Marek, J.; Korabik, M.; Mrozinski, J. Syntheses and Properties of Binuclear Copper(II) Mixed-Ligand Complexes Involving Thiodiglycolic Acid. *Polyhedron* **2003**, *22*, 411–418. [CrossRef]
51. Grirrane, A.; Pastor, A.; Galindo, A.; del Río, D.; Orlandini, A.; Mealli, C.; Ienco, A.; Caneschi, A.; Fernández Sanz, J. Supramolecular Interactions as Determining Factors of the Geometry of Metallic Building Blocks: Tetracarboxylate Dimanganese Species. *Angew. Chem. Int. Ed.* **2005**, *44*, 3429–3432. [CrossRef] [PubMed]
52. Neuman, N.I.; Burna, E.; Baggio, R.; Passeggi, M.C.G.; Rizzi, A.C.; Brondino, C.D. Transition from Isolated to Interacting Copper(II) Pairs in Extended Lattices Evaluated by Single Crystal EPR Spectroscopy. *Inorg. Chem. Front.* **2015**, *2*, 837–845. [CrossRef]
53. Grirrane, A.; Pastor, A.; Álvarez, E.; Mealli, C.; Ienco, A.; Galindo, A. Novel Results on Thiodiacetate Zinc(II) Complexes: Synthesis and Structure of [Zn(Tda)(Phen)]$_2$·5H$_2$O. *Inorg. Chem. Commun.* **2006**, *9*, 160–163. [CrossRef]
54. Sun, D.; Xu, M.-Z.; Liu, S.-S.; Yuan, S.; Lu, H.-F.; Feng, S.-Y.; Sun, D.-F. Eight Zn(II) Coordination Networks Based on Flexible 1,4-Di(1H-Imidazol-1-Yl)Butane and Different Dicarboxylates: Crystal Structures, Water Clusters, and Topologies. *Dalton Trans.* **2013**, *42*, 12324. [CrossRef]
55. Packiaraj, S.; Kanchana, P.; Pushpaveni, A.; Puschmann, H.; Govindarajan, S. Different Coordination Geometries of Lighter Lanthanates Driven by the Symmetry of Guanidines as Charge Compensators. *New J. Chem.* **2019**, *43*, 979–991. [CrossRef]
56. Malmborg, T.; Oskarsson, Å.; Rømming, C.; Gronowitz, S.; Koskikallio, J.; Swahn, C.-G. Structural Studies on the Rare Earth Carboxylates. 22. The Crystal Structure of Tetra-Aquo-Thiodiacetatoneodymium(III) Chloride. *Acta Chem. Scand.* **1973**, *27*, 2923–2929. [CrossRef]
57. Kepert, C.J.; Skelton, B.W.; White, A.H. Structural Systematics of Rare Earth Complexes. XXI Polymeric Sodium Bis(Thiodiglycolato)Neodymiate(III). *Aust. J. Chem.* **1999**, *52*, 617. [CrossRef]
58. Ghadermazi, M.; Olmstead, M.M.; Rostami, S.; Attar Gharamaleki, J. Poly[[Piperazine-1,4-Dium [Diaquatetrakis(µ-Sulfanediyldiacetato)Dicerate(III)]] Trihydrate]. *Acta Crystallogr. Sect. E Struct. Rep. Online* **2011**, *67*, m291–m292. [CrossRef]
59. Hou, X.; Li, D.; Wang, X.; Wang, J.; Ren, Y.; Zhang, M. Syntheses, Structures and Luminescence Properties of Ln-Coordination Polymers Based on Flexible Thiodiacetic Acid Ligand. *Chin. J. Chem.* **2009**, *27*, 1481–1486. [CrossRef]
60. Wen, H.-R.; Dong, P.-P.; Liang, F.-Y.; Liu, S.-J.; Xie, X.-R.; Tang, Y.-Z. A Family of 2D Lanthanide Complexes Based on Flexible Thiodiacetic Acid with Magnetocaloric or Ferromagnetic Properties. *Inorg. Chim. Acta* **2017**, *455*, 190–196. [CrossRef]
61. Wang, H.-S.; Bao, W.-J.; Ren, S.-B.; Chen, M.; Wang, K.; Xia, X.-H. Fluorescent Sulfur-Tagged Europium(III) Coordination Polymers for Monitoring Reactive Oxygen Species. *Anal. Chem.* **2015**, *87*, 6828–6833. [CrossRef] [PubMed]
62. Hosseinabadi, F.; Ghadermazi, M.; Taran, M.; Derikvand, Z. Synthesis, Crystal Structure, Spectroscopic, Thermal Analyses and Biological Properties of Novel F-Block Coordination Polymers Containing 2,2′-Thiodiacetic Acid and Piperazine. *Inorg. Chim. Acta* **2016**, *443*, 186–197. [CrossRef]
63. Zhang, Y.-Z.; Li, J.-R.; Gao, S.; Kou, H.-Z.; Sun, H.-L.; Wang, Z.-M. Two-Dimensional Rare Earth Coordination Polymers Involving Different Coordination Modes of Thiodiglycolic Acid. *Inorg. Chem. Commun.* **2002**, *5*, 28–31. [CrossRef]
64. Gans, P. GLEE, a New Computer Program for Glass Electrode Calibration. *Talanta* **2000**, *51*, 33–37. [CrossRef] [PubMed]
65. Petit, L.D.; Powell, K.J. *Stability Constants Database, SC-Database for Windows 1997*; Academic Software: Lokeren, Belgium, 1997.
66. Klungness, G.D.; Byrne, R.H. Comparative Hydrolysis Behavior of the Rare Earths and Yttrium: The Influence of Temperature and Ionic Strength. *Polyhedron* **2000**, *19*, 99–107. [CrossRef]
67. Martínez, S.; Igoa, F.; Carrera, I.; Seoane, G.; Veiga, N.; De Camargo, A.S.S.; Kremer, C.; Torres, J. A Zn(II) Luminescent Complex with a Schiff Base Ligand: Solution, Computational and Solid State Studies. *J. Coord. Chem.* **2018**, *71*, 874–889. [CrossRef]
68. *CrysAlisPro*, Version 2021; Rigaku Oxford Diffraction: Oxford, UK, 2021.
69. Sheldrick, G.M. *SHELXT*–Integrated Space-Group and Crystal-Structure Determination. *Acta Crystallographica. Sect. A Found. Adv.* **2015**, *71*, 3–8. [CrossRef]
70. Sheldrick, G.M. Crystal Structure Refinement with *SHELXL*. *Acta Crystallogr. Sect. C Struct. Chem.* **2015**, *71*, 3–8. [CrossRef]
71. Spek, A.L. Structure Validation in Chemical Crystallography. *Acta Crystallogr. D Biol. Crystallogr.* **2009**, *65*, 148–155. [CrossRef]
72. Dolomanov, O.V.; Bourhis, L.J.; Gildea, R.J.; Howard, J.A.K.; Puschmann, H. *OLEX2*: A Complete Structure Solution, Refinement and Analysis Program. *J. Appl. Crystallogr.* **2009**, *42*, 339–341. [CrossRef]
73. Bessen, N.P.; Popov, I.A.; Heathman, C.R.; Grimes, T.S.; Zalupski, P.R.; Moreau, L.M.; Smith, K.F.; Booth, C.H.; Abergel, R.J.; Batista, E.R.; et al. Complexation of Lanthanides and Heavy Actinides with Aqueous Sulfur-Donating Ligands. *Inorg. Chem.* **2021**, *60*, 6125–6134. [CrossRef] [PubMed]
74. Thakur, P.; Pathak, P.N.; Gedris, T.; Choppin, G.R. Complexation of Eu(III), Am(III) and Cm(III) with Dicarboxylates: Thermodynamics and Structural Aspects of the Binary and Ternary Complexes. *J. Solut. Chem.* **2009**, *38*, 265–287. [CrossRef]

75. Torres, J.; Peluffo, F.; Domínguez, S.; Mederos, A.; Arrieta, J.M.; Castiglioni, J.; Lloret, F.; Kremer, C. 2,2′-Oxydiacetato-Bridged Complexes Containing Sm(III) and Bivalent Cations. Synthesis, Structure, Magnetic Properties and Chemical Speciation. *J. Mol. Struct.* **2006**, *825*, 60–69. [CrossRef]
76. Grenthe, I.; Tobiasson, I.; Theander, O.; Hatanaka, A.; Munch-Petersen, J. Thermodynamic Properties of Rare Earth Complexes. I. Stability Constants for the Rare Earth Diglycolate Complexes. *Acta Chem. Scand.* **1963**, *17*, 2101–2112. [CrossRef]
77. Grenthe, I.; Gårdhammar, G.; Søtofte, I.; Beronius, P.; Engebretsen, J.E.; Ehrenberg, L. Thermodynamic Properties of Rare Earth Complexes. X. Complex Formation in Aqueous Solution of Eu(III) and Iminodiacetic Acid. *Acta Chem. Scand.* **1971**, *25*, 1401–1407. [CrossRef]
78. Chung, D.Y.; Lee, E.H.; Kimura, T. Laser-Induced Luminescence Study of Samarium(III) Thiodiglycolate Complexes. *Bull. Korean Chem. Soc.* **2003**, *24*, 1396–1398.
79. Lis, S.; Choppin, G.R. Luminescence Study of Europium(III) Complexes with Several Dicarboxylic Acids in Aqueous Solution. *J. Alloys Compd.* **1995**, *225*, 257–260. [CrossRef]
80. Dellien, I.; Grenthe, I.; Hessler, G. Thermodynamic Properties of Rare Earth Complexes XVIII. Free Energy, Enthalpy and Entropy Changes for the Formation of Some Lanthanoid Thiodiacetate and Hydrogen Thiodiacetate Complexes. *Acta Chem. Scandinava* **1973**, *27*, 2431–2440. [CrossRef]
81. Puentes, R.; Torres, J.; Kremer, C.; Cano, J.; Lloret, F.; Capucci, D.; Bacchi, A. Mononuclear and Polynuclear Complexes Ligated by an Iminodiacetic Acid Derivative: Synthesis, Structure, Solution Studies and Magnetic Properties. *Dalton Trans.* **2016**, *45*, 5356–5373. [CrossRef]
82. Casanova, D.; Llunell, M.; Alemany, P.; Alvarez, S. The Rich Stereochemistry of Eight-Vertex Polyhedra: A Continuous Shape Measures Study. *Chem.-Eur. J.* **2005**, *11*, 1479–1494. [CrossRef]
83. Llunell, M.; Casanova, D.; Cirera, J.; Alemany, P.; Alvarez, S. *SHAPE*, Version 1.1; Universitat de Barcelona: Barcelona, Spain, 2003.
84. Bünzli, J.-C.G. On the Design of Highly Luminescent Lanthanide Complexes. *Coord. Chem. Rev.* **2015**, *293–294*, 19–47. [CrossRef]
85. Thomsen, M.S.; Nawrocki, P.R.; Kofod, N.; Sørensen, T.J. Seven Europium(III) Complexes in Solution–The Importance of Reporting Data When Investigating Luminescence Spectra and Electronic Structure. *Eur. J. Inorg. Chem.* **2022**, *2022*, e202200334. [CrossRef]
86. Binnemans, K. Interpretation of Europium(III) Spectra. *Coord. Chem. Rev.* **2015**, *295*, 1–45. [CrossRef]

Disclaimer/Publisher's Note: The statements, opinions and data contained in all publications are solely those of the individual author(s) and contributor(s) and not of MDPI and/or the editor(s). MDPI and/or the editor(s) disclaim responsibility for any injury to people or property resulting from any ideas, methods, instructions or products referred to in the content.

Article

Functionalization and Coordination Effects on the Structural Chemistry of Pendant Arm Derivatives of 1,4,7-trithia-10-aza-cyclododecane ([12]aneNS$_3$)

Claudia Caltagirone [1], Maria Carla Aragoni [1], Massimiliano Arca [1], Alexander John Blake [2,*], Francesco Demartin [3], Alessandra Garau [1], Enrico Podda [4], Alexandra Pop [5], Vito Lippolis [1,*] and Cristian Silvestru [5]

[1] Dipartimento di Scienze Chimiche e Geologiche, Università degli Studi di Cagliari, S.S. 554 Bivio per Sestu, 09042 Monserrato (CA), Italy
[2] School of Chemistry, University of Nottingham, University Park, Nottingham NG7 2RD, UK
[3] Dipartimento di Chimica, Università degli Studi di Milano, 20133 Milano, Italy
[4] Centro Servizi di Ateno per la Ricerca-CeSAR, Università degli Studi di Cagliari, S.S. 554 Bivio per Sestu, 09042 Monserrato (CA), Italy
[5] Supramolecular Organic and Organometallic Chemistry Centre, Department of Chemistry, Faculty of Chemistry and Chemical Engineering, Babes-Bolyai University, Str. Arany Janos 11, 400028 Cluj-Napoca, Romania
* Correspondence: alexander.blake@ntlworld.com (A.J.B.); lippolis@unica.it (V.L.)

Abstract: The effect of different pendant arms on the structural chemistry of the 1,4,7-trithia-10-aza-cyclododecane ([12]aneNS$_3$) macrocycle is discussed in relation to the coordination chemistry of all known functionalized derivatives of [12]aneNS$_3$, which have been structurally characterized.

Keywords: 1,4,7-trithia-10-aza-cyclododecane; macrocyclic ligands; pendant arms; conformation

1. Introduction

1,4,7-Trithia-10-aza-cyclododecane ([12]aneNS$_3$ (**L** in Figure 1) is a well-known 12-membered mixed thia-aza macrocycle featuring a NS$_3$ donor set suitable for coordination to soft metal ions [1–15]. It has been extensively functionalized at the secondary nitrogen atom to afford pendant arm derivatives for different purposes: fluorescent materials and chemosensors for heavy metal ions [3,4,9–11], highly selective heteroditopic ionophores for simultaneous binding, extraction and transport of both the cationic and anionic moieties of toxic and/or precious transition metal salts [12,13], crystal engineering in the preparation of coordination polymers [5,8,12], tuning of the coordination environment around soft metal ions [2,6,7,15], and development of new synthetic methods of mixed-donor thiacrown ethers [1,14]. All structurally characterized pendant arm derivatives of [12]aneNS$_3$ (**L**) are reported in Figure 1, including both metal-coordinated ligands (drawn in black) and free ligands (drawn in red). Of note, no X-ray crystal structure of **L** is known, and **L7** is the only derivative of [12]aneNS$_3$ for which the X-ray crystal structure is known together with some of its transition metal coordination compounds. In this paper, we report the X-ray crystal structure of **L12** and the synthesis and structural characterization of two new **L** derivatives, i.e., **L20** and **L21**·CHCl$_3$ (Figure 2). The two new derivatives of **L** were chosen to favor intermolecular interactions involving the functional groups in the pendant arm and the donor atoms from the macrocyclic moiety. In particular, in the case of **L20**, we wanted to explore the possibility of intermolecular halogen bonds (XBs) formation involving the S donor atoms of the macrocyclic moiety and the effects on its conformation. The three new crystal structures are discussed and compared with those already known in order to identify particular trends in the conformational changes in the macrocyclic moiety upon

coordination with metal ions, and the geometrical effects induced by pendant arms bearing additional functionalities.

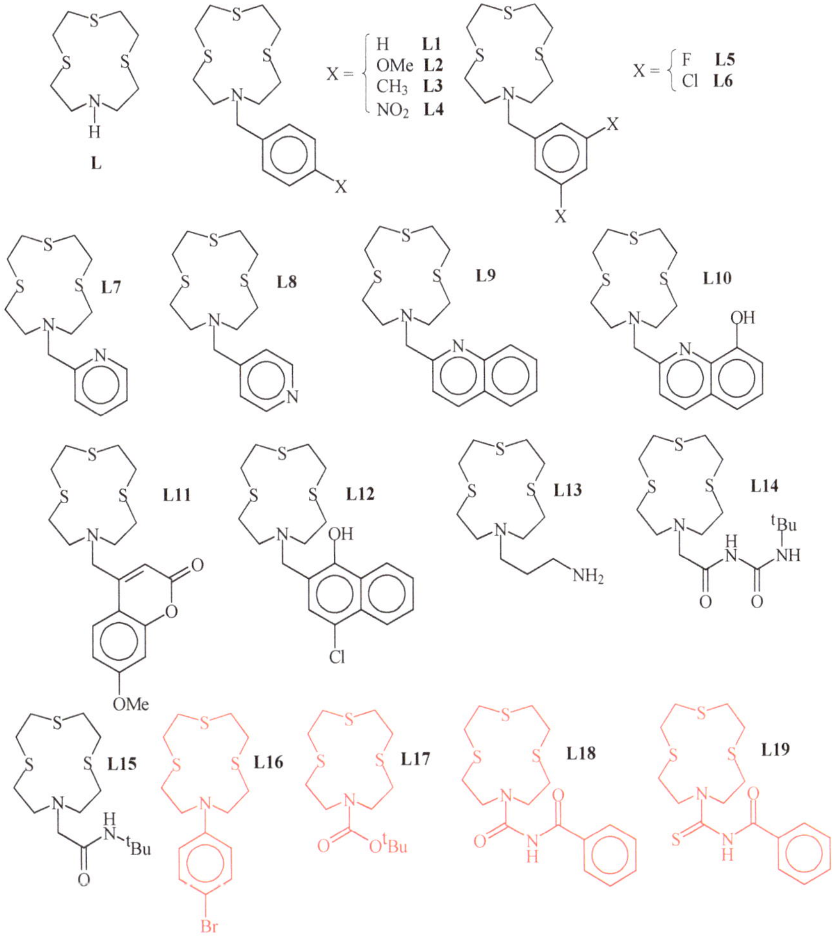

Figure 1. Summary of the crystallographically characterized [12]aneNS$_3$ (**L**) derivatives reported in the Cambridge Structural Database (CSD accessed on 13 January 2023) either as metal-coordinated ligands (drawn in black) or as free ligands (drawn in red). [LH]Br·2H$_2$O [1]; [Cd(**L**)(NO$_3$)$_2$] [2]; [{Tl(**L**)}$_2${Au(C$_6$Cl$_5$)$_2$}][Au(C$_6$Cl$_5$)$_2$], [{Tl(**L**)}{Au(C$_6$F$_5$)$_2$}]$_2$ [3]; [{Ag(**L**)}{Au(C$_6$F$_5$)$_2$}]$_2$ [4]; {[Ag(**L1**)](OTf)·MeCN}$_\infty$, {[Ag(**L2**)](PF$_6$)·MeCN}$_\infty$, {[Ag(**L3**)](OTf)·MeCN}$_\infty$, [Ag$_3$(**L4**)$_3$](OTf)$_3$·4MeCN, {[Ag(**L5**)](OTf)·Me$_2$CO}$_\infty$, [Ag$_3$(**L6**)$_3$](OTf)$_3$·2MeOH·H$_2$O [5]; **L7**, [Hg(**L7**)][HgCl$_4$], [Cd(**L7**)I][CdI$_4$] [6]; [Mo(CO)$_3$(**L7**)] [7]; [Ag$_2$(**L7**)$_2$](OTf)$_2$·2MeCN, [Ag$_2$(**L8**)$_2$](OTf)$_2$·2MeCN [8]; [Ag(**L9**)][Au(C$_6$F$_5$)$_2$]·THF, [{Au(C$_6$F$_5$)$_2$}$_2$Ag$_2${Au(C$_6$F$_5$)$_2$}$_2${Ag(**L9**)}$_2$]·2THF [4]; [Zn(**L10**)](ClO$_4$)$_2$ [9]; [Hg(**L11**$_{-H}$)](NO$_3$) [10]; [Zn(**L12**)$_2$H$_2$O](BF$_4$)$_2$ [9]; [Hg(**L13**)](ClO$_4$)$_2$ [11]; [CuCl(**L14**)][CuCl$_4$], {[Ag(**L14**)](NO$_3$)}$_\infty$ [12]; {[Ag(**L15**)](NO$_3$)}$_\infty$ [13]; **L16** [14]; **L17** [15]; **L18**, **L19** [13].

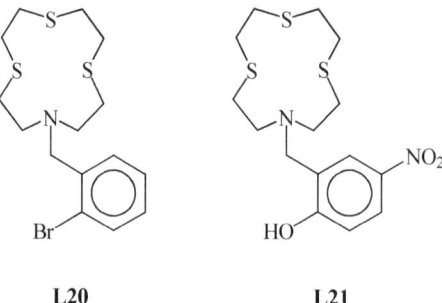

Figure 2. New pendant arm derivatives of **L** discussed in this paper.

2. Materials and Methods

2.1. Materials and Methods

1,4,7-Trithia-10-aza-cyclododecane ([12]aneNS$_3$) [11] and **L12** were synthesized according to the procedure reported in the literature [9]. Reagents and solvents were used that were purchased from Aldrich. Elemental analyses were performed with an EA1108 CHNS-O Fisons instrument (T = 1000 °C). The ^{1}H- and ^{13}C-NMR spectra were determined on a Bruker Avance 600 MHz spectrometer. The mass spectra were recorded on a triple quadrupole QqQ Varian 310-MS mass spectrometer by using the atmospheric pressure ESI technique. All sample solutions were infused into the ESI source with a programmable syringe pump (1.50 mL/h constant flow rate). A dwell time of 14 s was used, and the spectra were accumulated for at least 10 min to increase the signal-to-noise ratio. The mass spectra were recorded in the m/z 100–1000 range. The following scan parameters were chosen [16]: needle voltage 3500 V, shield 800 V, source temperature 60 °C, drying gas pressure 20 psi, nebulizing gas pressure 20 psi, detector voltage 1450 V, and drying gas temperature 110 °C. The isotopic patterns of the measured peaks in the mass spectra were analyzed using the mMass 5.5.0 software package [17]. All mass values were indicated as monoisotopic masses, which were computed as the sum of the masses of the primary isotope of each atom in the molecule (note that the monoisotopic mass may differ from the nominal molecular mass).

2.2. X-ray Diffraction Analyses

Only special features are noted here. Crystallographic data are reported in Table S1 in the Supplementary Materials. Single-crystal X-ray diffraction (SC-XRD) data for compound **L12** were collected at 150 K on a Bruker SMART 1000 diffractometer using MoKα radiation. The data were indexed and processed using Bruker SAINT and SMART [18,19]. The structure was solved by the SIR92 [20] solution program using direct methods. SC-XRD data for compound **L20** were collected at 293 K on an Agilent SuperNova diffractometer equipped with an Eos detector using MoKα radiation. The data were indexed and processed using CrysalisPro [21] and the structure was solved by SHELXT [22] using dual-space methods. SC-XRD data for the compound **L21**·CHCl$_3$ were collected at 293 K on a Bruker APEX II CCD diffractometer using MoKα radiation. The data were indexed and processed using Bruker SAINT [18]. The structure was solved by SIR92 [20] using direct methods. The models were refined with ShelXL [23] using full-matrix least-squares minimization on F^2. All non-hydrogen atoms were anisotropically refined. Hydrogen atom positions were geometrically calculated and refined using the riding model. Olex2, version 1.5 [24], was used as the graphical interface. For the compound **L21**·CHCl$_3$, the solvent molecule was disordered and modelled over two sites with statistically the same occupancy [0.49(1) and 0.51(1), respectively]. Complete crystallographic data for the structure of compounds **L12**, **L20**, and **L21**·CHCl$_3$ were deposited in CIF format at the Cambridge Crystallographic Data Center (CCDC) with deposition numbers 2248158-2248160, respectively.

*2.3. Synthesis of **L20***

A mixture of 1,4,7-trithia-10-aza-cyclododecane (0.178 g, 0.790 mmol), 2-bromobenzyl bromide (0.199 g 0.790 mmol), and K$_2$CO$_3$ (0.37 g, 2.6 mmol) in dry MeCN (50 mL) was heated to reflux for 96 h under a dry N$_2$ atmosphere. The solid was filtered off and the solvent was removed under reduced pressure. The residue was dissolved in dichloromethane (DCM) and washed with distilled water. The organic phase was dried over Na$_2$SO$_4$ and the solvent was removed under reduced pressure. The residue was purified by flash chromatography (silica) using DCM/MeOH (10:0.3, v/v) as eluent to give a pale-yellow solid (0.187 mg, 0.63 mmol, 80% yield). **Elemental analysis, Found (Calcd. for C$_{15}$H$_{22}$BrNS$_3$):** C, 45.68 (45.91); H, 5.58 (5.65); N, 3.71 (3.57); S, 24.42 (24.51)%. M.p. 108 °C; ^{1}H-NMR (CDCl$_3$, 600 MHz): δ_H (ppm) 2.65–2.69 (m, 4H), 2.78–2.82 (m, 12H), 3.71 (s, 2H, NCH$_2$Ar), 7.12 (t, J = 7.1 Hz, 1H), 7.29 (t, J = 7.3 Hz, 1H), and 7.54 (d, J = 7.5 Hz, 2H); ^{13}C{^{1}H}-NMR (CDCl$_3$, 150 MHz): δ_C (ppm) 25.90, 27.86, 28.49 (CH$_2$S), 51.86 (CH$_2$N), 59.55 (NCH$_2$Ar), 124.38, 127.41, 129.70, 130.69, 132.85, and 137.88 (aromatic carbons); and ESI-MS(+): m/z = 393 [M+H]$^+$. The crystals of **L20** suitable for an X-ray diffraction analysis were grown from a DCM solution by diffusion of Et$_2$O vapor.

*2.4. Synthesis of **L21***

A weighted amount of 2-hydroxy-5-nitrobenzyl bromide (0.170 g, 0.740 mmol) was added to a solution of 1,4,7-trithia-10-aza-cyclododecane (0.150 g, 0.670 mmol) and K$_2$CO$_3$ (0.460 g, 3.33 mmol) in dry MeCN (20 mL). The reaction mixture was heated under nitrogen at 80 °C for 24 h and then for 24 h at room temperature. The solid was filtered off, and the solvent was removed under reduced pressure to give a yellow solid (0.170 mg, 0.45 mmol, 68% yield). **Elemental analysis, Found (Calcd. for C$_{15}$H$_{22}$N$_2$O$_3$S$_3$):** C, 48.51 (48.10);H, 6.12 (5.92); N, 7.48 (7.48); S, 25.91 (25.68)%. M.p. 175 °C; ^{1}H-NMR (CDCl$_3$/CD$_3$CN, 600 MHz): δ_H (ppm) 2.68 (m, 4H), 2.76 (m, 8H), 2.81 (m, 4H), 3.71 (s, 2H, NCH$_2$Ar), 6.56 (d, J = 9.2 Hz, 1H), 7.91 (dd, J = 3.0 Hz, 1H), and 8.05 (d, J = 2.8 Hz, 1H); ^{13}C{^{1}H}-NMR (CDCl$_3$/CD$_3$CN, 150 MHz): δ_C (ppm) 26.5, 29.2, 29.4(CH$_2$S), 52.5 (CH$_2$N), 56.1 (NCH$_2$Ar), 118.7, 126.2, 126.9, 127.4, 136.7, and 172.6 (aromatic carbons); and ESI-MS(+): m/z = 375 [M+H]$^+$. The crystals of **L21**·CHCl$_3$ suitable for an X-ray diffraction analysis were grown from a CHCl$_3$ solution.

3. Results and Discussion

The X-ray crystal structure of **L12** is characterized by the formation of dimeric arrangements via N···H–O hydrogen bonds involving the –OH group and the aromatic N atom from the pendant arms of two symmetry-related ligand units (Figure 3). π-π Stacking interactions between the quinoline moieties of adjacent slipped dimers (centroid-centroid distance: 3.77 Å) generate a staircase architecture along the *b*-axis as shown in Figure S1 in the Supplementary Materials.

In **L12**, the macrocyclic unit adopts a square-shaped conformation with the four donor atoms occupying the corners. A useful convention to describe the different conformations that an aliphatic macrocycle can adopt is that of Dale [25], and it is based on the sequence of torsion angles around the macrocycle ring, which normally are classified as either *anti* (≅180°) or *gauche* (≅60°). According to the Dale convention, a generic conformation is determined by the number of bonds between each corner in the macrocycle (a corner is the atom at which two *gauche* torsion angles intersect). Considering the sequence of the torsion angles in **L12** (Table 1), four corners in the [12]aneNS$_3$ unit are present at the N and S donor atoms. Each pair of corners is separated by one −N−CH$_2$−CH$_2$−S− or −S−CH$_2$−CH$_2$−S− side-moiety unit featuring an *anti*-torsion angle at the C−C bond. Therefore, the conformation adopted by the macrocyclic unit in **L12** is [3333], according to the Dale convention, with the lone pairs (LPs) of electrons on the donor atoms in *exo*-dentate positions (i.e., pointing out of the ring cavity).

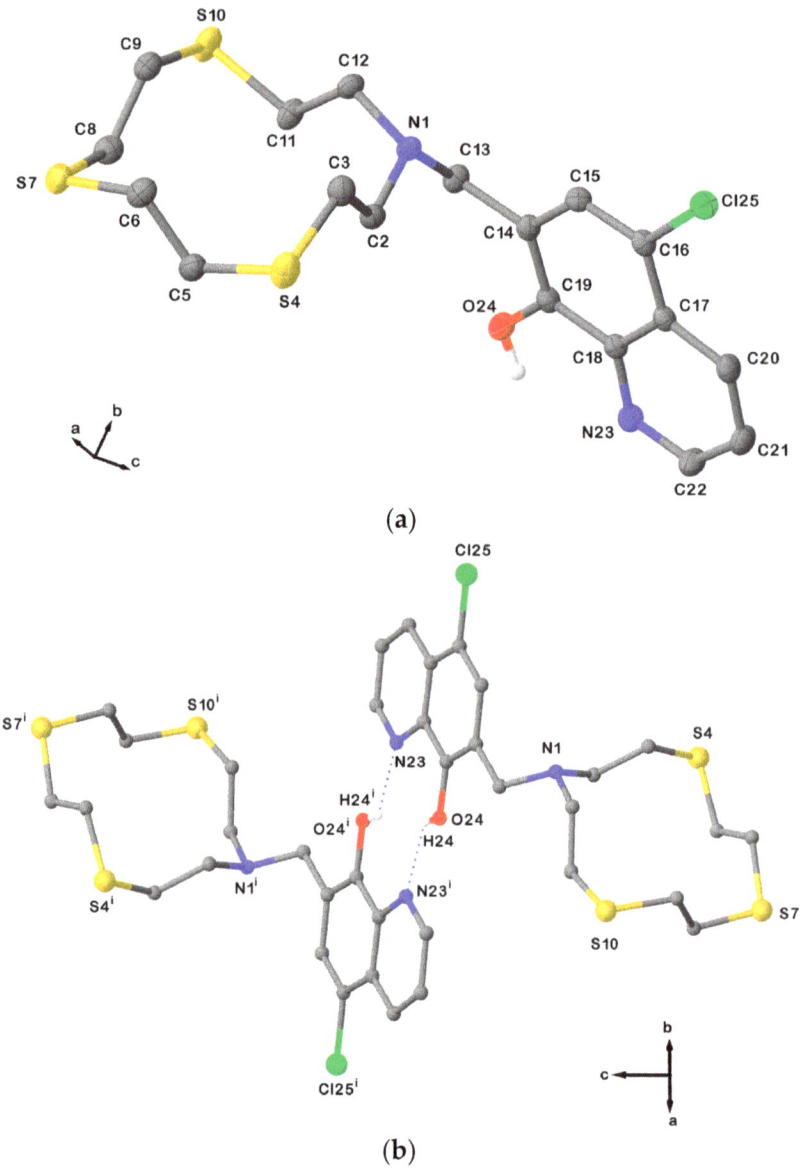

Figure 3. (a) View of the asymmetric unit of **L12** with the atom-labelling scheme adopted. Hydrogen atoms are omitted for clarity. Displacement ellipsoids are drawn at 50% probability level; (b) H-bonded dimeric arrangement found in **L12** with the numbering scheme adopted. Only interacting hydrogen atoms are shown for clarity. N23···O24^i = 2.8514(6) Å; O24^i−H24^i···N23 = 135°; and symmetry operation: i = −x, −y, 2−z. For selected bond lengths (Å) and angles (°), see Table S2 in the Supplementary Materials.

Table 1. Torsion angles (°) for the [12]aneNS$_3$ moieties in **L12**, **L20,** and **L21**·CHCl$_3$ with the Dale conformations in bracket squares (*gauche* torsion angles in red).

	L12 [3333]	L20 [11334]	L21·CHCl$_3$ [3333]
N1−C2−C3−S4	−172.90(14)	−66.0(3)	172.32(9)
C2−C3−S4−C5	77.12(18)	88.4(3)	−74.74(12)
C3−S4−C5−C6	75.95(19)	74.0(3)	−72.83(12)
S4−C5−C6−S7	−168.74(13)	175.40(17)	170.11(8)
C5−C6−S7−C8	65.4(2)	59.4(3)	−70.10(12)
C6−S7−C8−C9	67.8(2)	63.3(3)	−68.69(12)
S7−C8−C9−S10	−173.99(13)	−166.01(19)	173.80(8)
C8−C9−S10−C11	73.4(2)	68.5(3)	−70.48(12)
C9−S10−C11−C12	75.6(2)	66.8(3)	−75.57(12)
S10−C11−C12−N1	−165.24(17)	−155.5(2)	166.58(10)
C11−C12−N1−C2	64.8(3)	166.2(3)	−67.18(6)
C12−N1−C2−C3	69.1(2)	−73.8(4)	−69.40(5)

This type of conformation with *exo*-dentate donor atoms positioned at the corners is very common in medium and large aliphatic thioether macrocycles, as demonstrated by diffraction studies [26–28]: an *anti*-torsion angle at the C−C bond in the −S−CH$_2$−CH$_2$−S− units is preferred as a consequence of the repulsive 1,4-interaction between the sulfur LPs in a *gauche* arrangement [29]. On the other hand, a *gauche* torsion angle at the C−S bond in a −CH$_2$−CH$_2$−S−CH$_2$− sequence is not disfavored by the 1,4-interactions of the methylene protons, which lie beyond the sum of the corresponding van der Waals radii [30]. An opposite trend is observed in oxa-macrocycles due to the different C−O bond length and covalent radius of the donor atom [31].

In general, and in the specific case of **L12**, metal complexation or the involvement of one of the donor atoms of the [12]aneNS$_3$ unit in interactions with functionalities present in the pendant arms can strongly affect the conformation adopted by the macrocyclic unit in the solid state.

In fact, the X-ray crystal structure of **L20** is also characterized by dimeric arrangements via C−Br···S halogen bonds (XBs) involving one of the S donors from the macrocyclic unit as an XB acceptor, and the Br atom on the benzylic pendant arm as an XB donor (Figure 4).

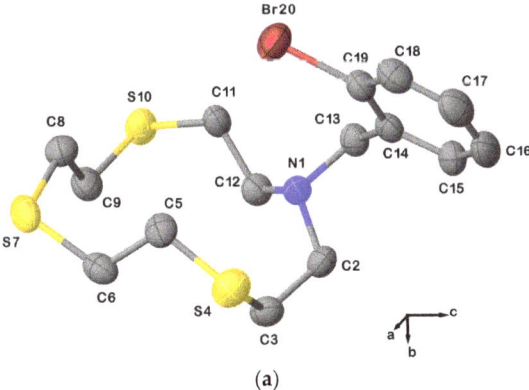

(a)

Figure 4. *Cont.*

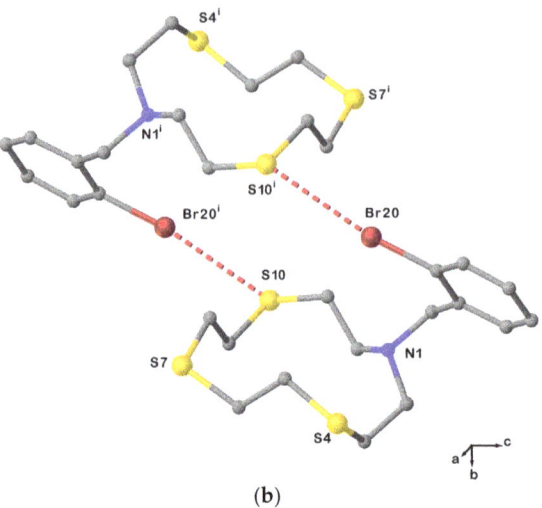

(b)

Figure 4. (a) View of the asymmetric unit of **L20** with the atom-labelling scheme adopted. Hydrogen atoms have been omitted for clarity. Displacement ellipsoids are drawn at 50% probability level; (b) X-bonded dimeric arrangement found in **L20** with the numbering scheme adopted. S10···Br20^i = 3.5030(3) Å; C19−Br20···S10^i = 163.6(1)°; and symmetry operation: i = −x, 1−y, −z. For selected bond lengths (Å) and angles (°), see Table S3 in the Supplementary Materials.

Due to this intermolecular interaction, the macrocyclic unit assumes a [11334] conformation as five corners can be identified in the cyclic structure (three consecutive at the atoms C2, C3, and S4, and the other two at S7 and S10, Table 1). Compared to **L12**, *anti*-torsion angles in **L20** remain at the C−C bonds in the sequences S4−C5−C6−S7, S7−C8−C9−S10, and S10−C11−C12−N1 (Table 1), while the torsion angles at the C2−C3 and C12−N1 bonds in the sequences N1−C2−C3−S4 and C11−C12−N1−C2 adopt a *gauche* and an *anti*-conformation, respectively (vice versa in the case of **L12**, Table 1). Presumably, the pulling effect of the bromine atom on the sulfur atom S10 in forming the XB causes the torsion angle at one of the C−N to change from a *gauche* to an *anti*-conformation, and the torsion angle at the C2−C3 bond to do the exact opposite.

This does not happen in the structure of **L21**·CHCl$_3$, in which it is the N atom from the macrocyclic unit that is involved in a weak intramolecular interaction with a functional group in the pendant arm. In fact, the conformation adopted by the [12]aneNS$_3$ unit in **L21**·CHCl$_3$ is again square-shaped [3333] with the *exo*-dentate donor atoms at the corners and −N−CH$_2$−CH$_2$−S− or −S−CH$_2$−CH$_2$−S− sequences with *anti*-torsion angles at the C−C bonds at the sides (Table 1). In this compound, an intramolecular HB is formed between the hydroxyl −OH function and the tertiary N donor from the macrocyclic unit (Figure 5a), which keeps the *exo*-dentate orientation of the LP in the HB interaction. Furthermore, dimeric arrangements are determined by C−H···O HBs involving one oxygen atom from the nitro group and one benzylic hydrogen atom (Figure 5b).

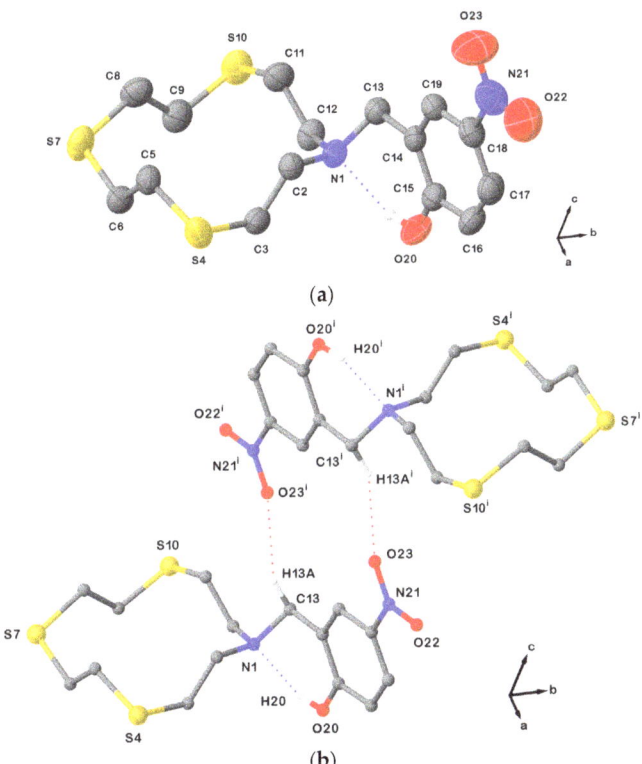

Figure 5. (a) View of the asymmetric unit in **L21**·CHCl$_3$ with the atom-labelling scheme adopted. Displacement ellipsoids are drawn at 50% probability level. Hydrogen atoms and the co-crystallized CHCl$_3$ molecule are omitted for clarity; (b) H-bonded dimeric arrangement found in **L21**·CHCl$_3$. Only interacting hydrogen atoms are shown for clarity. N1···O20 = 2.612(3); N1···H20 = 1.90(3) Å; O20−H20···N1 = 150(3)°; C13−O23^i = 3.354(3) Å; and symmetry operation: i = 1−x, 2−y, 2−z. For selected bond lengths (Å) and angles (°), see Table S4 in the Supplementary Materials.

A survey in the CSD (accessed on 13 January 2023) reveals a total of 31 X-ray crystal structures of [12]aneNS$_3$ (**L**) derivatives bearing functionalized pendant arms being reported either as free ligands or as metal complexes. The comparative analysis of the conformations assumed by the macrocyclic framework in the crystallographically independent 43 units (Table S5 in the Supplementary Materials) in terms of Dale notation shows that all reported free ligands (8 independent units) feature the [3333] conformation with the *exo*-dentate donor atoms at the corners (black asterisks in Figure 6a) which is in line with what is found in the structures of **L12** and **L21**·CHCl$_3$. The only exception is represented by the structure of **L20** mentioned above, and by the conformation assumed by the macrocycle in its bromide salt [HL]Br·2H$_2$O [1]. In this compound, the macrocycle is protonated at the secondary nitrogen atom and it assumes an unusual [84] (Table S5) conformation with only two corners, both at the carbon atoms, due to the intramolecular N−H···S HB involving the sulfur atom in front of the protonated N donor.

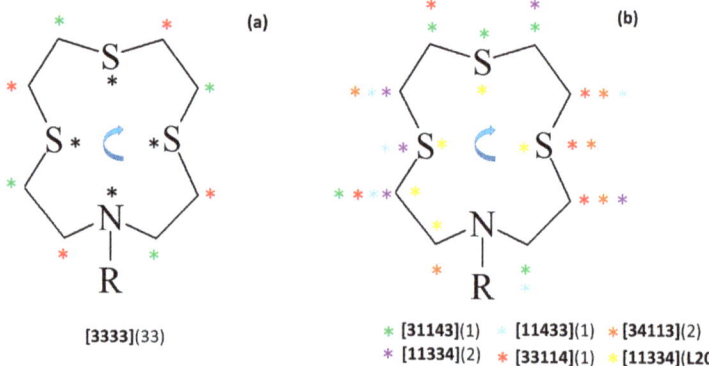

Figure 6. Schematic representation of the conformations adopted by the [12]aneNS$_3$ (**L**) macrocyclic moiety in the 31 X-ray crystal structures (43 independent units) retrieved from the CSD (Table S5 in the Supplementary Materials) featuring **L** derivatives (Figure 1) and their metal complexes. Asterisks indicate the corners in the various conformations according to the Dale notation. (**a**) Corner positions observed in [3333] conformations: black asterisks refer to the corner positions observed in free ligands, while red and green asterisks identify the corner positions observed in ligands coordinated to metal centers, and numbers in round brackets represent the overall number of independent units having the [3333] conformation; (**b**) corner positions in the alternative conformations observed in metal complexes of **L** derivatives, with each type of conformation identified by a different color of asterisk. The [84] conformation is not represented. To determine the number of bonds between each pair of corners and thereby assign the Dale notation to each conformation observed, start from the first corner to the left of the N atom as shown in the picture and proceed clockwise. This method was chosen to best illustrate the differences between conformations. The Dale notation is conventionally defined with the low numbers (i.e., the shortest edges) specified first, disregarding clockwise/anti-clockwise direction and crystallographic symmetries. In this way, all the conformations in Figure 6b would be assigned as [11334].

Among the remaining structures in which the macrocyclic moiety is involved in metal coordination, 25 independent units still display the macrocyclic moiety in a [3333] conformation with the corners located on carbon atoms and the donor atoms on –CH$_2$–CH$_2$–E–CH$_2$– (red asterisks in Figure 6a) or –CH$_2$–E–CH$_2$–CH$_2$– (green asterisks in Figure 6a), counting clockwise from the first corner to the left of the N atom as shown in Figure 6) side edges of the square-shaped conformation (E = N or S). A small number of compounds feature a conformation of the metal-coordinated macrocyclic unit in which three corners are located on three consecutive atoms including a S atom in the middle (purple, cyan, green, red, and brown asterisks in Figure 6b, see Table S5). Only in the case of **L20** does the sequence of three consecutive corners feature a S atom at the end of the three consecutive atoms involved (yellow asterisks in Figure 6b). Interestingly, no conformations with three consecutive corners involving N as one of the three consecutive atoms are known. Finally, only two structures are known in which the macrocyclic framework adopts a [84] conformation featuring only two corners (both at the same carbon atoms, Table S5), namely [HL]Br·2H$_2$O [1] and [CuCl(**L14**)]CuCl$_4$ [12].

4. Conclusions

The extent to which a macrocyclic ligand must pre-organize its conformation prior to complexation is one of the structural factors that can strongly affect the stability of the resulting metal complexes. For some macrocycles, the most stable conformation in the free state is not the most suitable to bind the metal ion and some changes are necessary to satisfy the stereochemical requirements of the complexation process. Herein, we have analyzed the conformation adopted in the solid state by the [12]aneNS$_3$ macrocyclic ring in all its functionalized pendant arm derivatives, both as free ligands and as metal complexes. The derivatives of this 12-membered tetradentate NS$_3$ macrocycle in their free state adopt a [3333] conformation, regardless of the nature of the functional groups in the pendant arms, with all 4 donor atoms at the corners of a square-shaped structure with the lone pairs in *exo*-dentate positions. This is also confirmed by the X-ray crystal structures of **L12** and **L21**·CHCl$_3$ presented herein. Therefore, upon complexation, a drastic conformational change is required to bring the lone pairs on the donor atoms into *endo*-dentate arrangements to maximize coordination to the metal ion. The most commonly observed conformation of the [12]aneNS$_3$ moiety bound to metal ions is still of the type [3333], but the donor atoms sacrifice their preferred corner positions and each becomes part of a "side" of the square-shaped structure, with each side made up of a sequence of *gauche- anti-gauche* torsion angles. In very few reported cases, *endo*-dentate arrangements of the donor atoms in the [12]aneNS$_3$ moiety are reached upon coordination by assuming conformations that are different from the [3333]. These conformations of the type [33114] (or related ones by cyclic permutations of the corner positions, see Figure 6 and Figure S2 in the Supplementary Materials) all feature three consecutive corners, i.e., three consecutive atoms at which two *gauche* torsion angles intersect, of which a S atom is always in the middle. The other three donor atoms (two S and one N) are not in corner positions. A similar conformation of the [12]aneNS$_3$ moiety, but with a S donor atom being a terminal corner in the sequence of three consecutive ones (Figure 6b), has been observed in the structure of the free ligand **L20**, and it is determined by an intermolecular C−Br···S halogen bond. In **L20**, the conformation adopted by the macrocyclic unit is [11334] and not [3333] as normally observed; the three S donor atoms are all in corner positions while the N donor atom is not and is characterized by an *endo*-dentate orientation of its lone pair. We conclude that this is the only case in which the functional group in the pendant arm affects the conformation assumed by the [12]aneNS$_3$ macrocyclic moiety due to a direct intermolecular interaction with one of the S donor atoms. Our results suggest that in order to have a conformation other than [3333] for the macrocyclic moiety of uncoordinated pendant arm derivatives of **L**, rather than considering intramolecular HBs involving the N atom and HB donors in the pendant arms (see the structure of **L21**·CHCl$_3$), it is more convenient to look for functionalities that can interact with the S donors of the macrocyclic moiety.

Supplementary Materials: The following supporting information can be downloaded at: https://www.mdpi.com/article/10.3390/cryst13040616/s1, Figure S1: Partial view along the [100] direction of the crystal packing in **L12** showing the slipped π-π stacking interactions of adjacent dimers: centroid-centroid distance = 3.77 Å, shift distance = 1.42 Å; Figure S2: Pictorial representation of the conformations assumed by the [12]aneNS$_3$ moiety in crystallographically characterized derivatives of the macrocycle with corners depicted as full circles; Table S1: Crystallographic data and structure refinement parameters for **L12**, **L20**, and **L21**·CHCl$_3$; Table S2: Selected bond lengths (Å) and angles (°) for **L12**; Table S3: Selected bond lengths (Å) and angles (°) for **L20**; Table S4: Selected bond lengths (Å) and angles (°) for **L21**·CHCl$_3$; Table S5: Torsion angles for the [12]aneNS$_3$ moiety in derivatives of the macrocycle reported in the Cambridge Structural Database (CSD). Dark green regions represent corners according to the Dale convention.

Author Contributions: Conceptualization, V.L. and A.J.B.; methodology, V.L.; validation, V.L. and A.J.B.; investigation, C.C., M.C.A., M.A., A.J.B., F.D., A.G., E.P. and A.P.; data curation, A.J.B., F.D., E.P. and V.L.; writing—original draft preparation, V.L.; writing—review and editing, C.C., M.C.A., M.A., A.J.B., F.D., A.G., E.P., A.P., V.L. and C.S.; supervision, V.L. All authors have read and agreed to the published version of the manuscript.

Funding: This research received no external funding.

Data Availability Statement: Crystallographic data in CIF format were deposited in the Cambridge Crystallographic Data Centre with deposition numbers CCDC: 2248158-2248160.

Acknowledgments: The University of Cagliari is acknowledged for financial support. CeSAR (Centro Servizi d'Ateneo per la Ricerca) of the University of Cagliari, Italy, is acknowledged for NMR measurements. A.J.B. thanks the University of Nottingham for access to facilities.

Conflicts of Interest: The authors declare no conflict of interest.

References

1. Glenny, M.W.; van de Water, L.G.A.; Vere, J.M.; Blake, A.J.; Wilson, C.; Driessen, W.L.; Reedijk, J.; Schröder, M. Improved synthetic methods to mixed-donor thiacrown ethers. *Polyhedron* **2006**, *25*, 599–612. [CrossRef]
2. Glenny, M.W.; van de Water, J.M.; Driessen, W.L.; Reedijk, J.; Blake, A.J.; Wilson, C.; Schröder, M. Conformational and stereochemical flexibility in cadmium(II) complexes of aza-thioether macrocycles. *Dalton Trans.* **2004**, *13*, 1953–1959. [CrossRef] [PubMed]
3. Donamaría, R.; Lippolis, V.; López-de-Luzuriaga, J.M.; Monge, M.; Nieddu, M.; Olmos, M.E. Tuning Au(I)···Tl(I) Interactions via Mixed Thia-Aza Macrocyclic Ligands: Effects in the Structural and Luminescence Properties. *Inorg. Chem.* **2017**, *56*, 12551–12563. [CrossRef]
4. Donamaría, R.; Lippolis, V.; López-de-Luzuriaga, J.M.; Monge, M.; Nieddu, M.; Olmos, M.E. Influence of the Number of Metallophilic Interactions and Structures on the Optical Properties of Heterometallic Au/Ag Complexes with Mixed-Donor Macrocyclic Ligands. *Inorg. Chem.* **2018**, *57*, 11099–11112. [CrossRef]
5. Habata, Y.; Noto, K.; Osaka, F. Substituents Effects on the Structures of Silver Complexes with Monoazatrithia-12-Crown-4 Ethers Bearing Substituted Aromatic Rings. *Inorg. Chem.* **2007**, *46*, 6529–6534. [CrossRef] [PubMed]
6. Hodorogea, A.M.; Silvestru, A.; Lippolis, V.; Pop, A. Group 12 metal complexes of mixed thia/aza and thia/oxa/aza macrocyclic ligands. *Polyhedron* **2022**, *216*, 115650. [CrossRef]
7. Ogawa, T.; Koike, K.; Matsumoto, J.; Kajita, Y.; Masuda, H. Synthesis and characterization of tricarbonyl-molybdenum complexes bearing monoaza-trithia-macrocyclic ligands. *Inorg. Chim. Acta* **2013**, *401*, 101–106. [CrossRef]
8. Habata, Y.; Osaka, F. Dimetallo[3.3]para- and metacyclophanes by self-assembly of pyridylmethyl armed-monoazatrithia- and monoazadithiaoxa-12-crown-4 ethers with Ag^+. *Dalton Trans.* **2006**, *15*, 1836–1841. [CrossRef]
9. Aragoni, M.C.; Arca, M.; Bencini, A.; Caltagirone, C.; Garau, A.; Isaia, F.; Light, M.E.; Lippolis, V.; Lodeiro, C.; Mameli, M.; et al. Zn^{2+}/Cd^{2+} optical discrimination by fluorescent chemosensors based on 8-hydroxyquinoline derivatives and sulfur-containing macrocyclic units. *Dalton Trans.* **2013**, *42*, 14516–14530. [CrossRef]
10. Bazzicalupi, C.; Caltagirone, C.; Cao, Z.; Chen, Q.; Di Natale, C.; Garau, A.; Lippolis, V.; Lvova, L.; Liu, H.; Lundström, I.; et al. Multimodal Use of New Coumarin-Based Fluorescent Chemosensors: Towards Highly Selective Optical Sensors for Hg^{2+} Probing. *Chem. Eur. J.* **2013**, *19*, 14639–14653. [CrossRef]
11. Aragoni, M.C.; Arca, M.; Bencini, A.; Blake, A.J.; Caltagirone, C.; Decortes, A.; Demartin, F.; Devillanova, F.A.; Faggi, E.; Dolci, L.S.; et al. Coordination Chemistry of *N*-aminopropyl pendant arm derivatives of mixed N/S-, and N/S/O-donor macrocycles, and construction of selective fluorimetric chemosensors for heavy metal ions. *Dalton Trans.* **2005**, *18*, 2994–3004. [CrossRef]
12. Love, J.B.; Vere, J.M.; Glenny, M.W.; Blake, A.J.; Schröder, M. Ditopic azathioethers macrocycles as hosts for transition metal salts. *Chem. Commun.* **2001**, *2678–2679*. [CrossRef]
13. Glenny, M.W.; Lacombe, M.; Love, J.B.; Blake, A.J.; Lindoy, L.F.; Luckay, R.C.; Gloe, K.; Antonioli, B.; Wilson, C.; Schröder, M. Design and synthesis of heteroditopic aza-thioether macrocycles for metal extraction. *New J. Chem.* **2006**, *30*, 1755–1767. [CrossRef]
14. Hoover, L.R.; Pryor, T.; Weitgenant, J.A.; Williams, P.E.; Storhoff, B.N.; Huffman, J.C. 4′ Diphenylphosphino and bromo derivatives of 10-phenyl-1,4,7-trithia-10-aza-cyclododecane 4-R-$C_6H_4N(CH_2CH_2S)_2CH_2CH_2SCH_2CH_2$. *Phosphorus Sulfur Silicon Relat. Elem.* **1997**, *122*, 155–166. [CrossRef]
15. Habata, Y.; Seo, J.; Otawa, S.; Osaka, F.; Noto, K.; Lee, S.S. Synthesis of diazahexathia-24-crown-8 derivatives structures of Ag^+ complexes. *Dalton Trans.* **2006**, *18*, 2202–2206. [CrossRef] [PubMed]
16. Masuri, S.; Cadoni, E.; Cabiddu, M.G.; Isaia, F.; Demuru, M.G.; Moraň, L.; Buček, D.; Vaňhara, P.; Havel, J.; Pivetta, T. The first copper(II) complex with 1,10-phenanthroline and salubrinal with interesting biochemical properties. *Metallomics* **2020**, *12*, 891–901. [CrossRef]
17. Strohalm, M.; Kavan, D.; Novak, P.; Volný, M.; Havlíček, V. mMass 3: Cross-platform Software Environment for Precise Analysis of Mass Spectrometric Data. *Anal. Chem.* **2010**, *82*, 4648–4651. [CrossRef]
18. *SAINT*, Version 6.36a; Bruker AXS Inc.: Fitchburg, WI, USA, 2002.
19. *SMART*, Version 5.624; Bruker AXS Inc.: Fitchburg, WI, USA, 2001.
20. Altomare, A.; Cascarano, G.; Giacovazzo, C.; Guagliardi, A.; Burla, M.C.; Polidori, G.; Camalli, M. SIR92—A program for automatic solution of crystal structures by direct methods. *J. Appl. Crystallogr.* **1994**, *27*, 435–436. [CrossRef]
21. *CrysAlisPro*, Version 1.171.37.35; Agilent Technologies: Oxfordshire, UK, 2014; Release 13-08-2014 CrysAlis171.NET.
22. Sheldrick, G.M. SHELXT—Integrated Space-Group and Crystal-Structure Determination. *Acta Crystallogr. Sect. A* **2015**, *71*, 3–8. [CrossRef]
23. Sheldrick, G.M. Crystal Structure Refinement with SHELXL. *Acta Crystallogr. Sect. C* **2015**, *71*, 3–8. [CrossRef]
24. Dolomanov, O.V.; Bourhis, L.J.; Gildea, R.J.; Howard, J.A.K.; Puschmann, H. OLEX2: A Complete Structure Solution, Refinement and Analysis Program. *J. Appl. Crystallogr.* **2009**, *42*, 339–341. [CrossRef]
25. Dale, J. Exploratory Calculations of Medium and Large Rings. *Acta Chem. Scand.* **1973**, *27*, 1115–1129. [CrossRef]
26. Simone, R.E.; Glick, M.D. Structures of the macrocyclic polythiaether 1,4,8,11-tetrathiacyclotetradecane and implications for transition-metal chemistry. *J. Am. Chem. Soc.* **1976**, *98*, 762–767.

27. Riley, D.P.; Oliver, J.D. Synthesis and crystal structure of a novel rhodium(I) complex of an inside-out hexathia crown ether. *Inorg. Chem.* **1983**, *22*, 3361–3363. [CrossRef]
28. Blake, A.J.; Gould, R.O.; Halcrow, M.A.; Schröder, M. Conformational studies on [16]aneS$_4$. Structures of α- and β-[16]aneS$_4$ ([16]aneS$_4$ = 1,5,9,13-tetrathiacyclohexadecane). *Acta Crystallogr. Sect. B* **1993**, *49*, 773–779. [CrossRef]
29. Wolf, R.E.; Hartman, J.-A.R.; Storey, J.M.E.; Foxman, B.M.; Cooper, S.R. Crown thioether chemistry: Structural and conformational studies of tetrathia-12-crown-4, pentathia-15-crown-5, and hexathia-18-crown-6. Implications for ligand design. *J. Am. Chem. Soc.* **1987**, *109*, 4328–4335. [CrossRef]
30. Bovil, M.J.; Chadwick, D.J.; Sutherland, I.O. Molecular mechanics calculations for ethers. The conformations of some crown ethers and the structure of the complex of 18-crown-6 with benzylammonium thiocyanate. *J. Chem. Soc. Perkin Trans.* **1980**, *2*, 1529–1543. [CrossRef]
31. Mark, J.E.; Flory, P.J. The Configuration of the Polyoxyethylene Chain. *J. Am. Chem. Soc.* **1965**, *87*, 1415–1423. [CrossRef]

Disclaimer/Publisher's Note: The statements, opinions and data contained in all publications are solely those of the individual author(s) and contributor(s) and not of MDPI and/or the editor(s). MDPI and/or the editor(s) disclaim responsibility for any injury to people or property resulting from any ideas, methods, instructions or products referred to in the content.

Article

Anion Coordination into Ligand Clefts

Matteo Savastano [1,2,*], Carlotta Cappanni [1], Carla Bazzicalupi [1,*], Cristiana Lofrumento [1] and Antonio Bianchi [1]

[1] Department of Chemistry "Ugo Schiff", University of Florence, Via della Lastruccia, 3-13, 50019 Sesto Fiorentino, Italy; carlotta.cappanni@stud.unifi.it (C.C.); cristiana.lofrumento@unifi.it (C.L.); antonio.bianchi@unifi.it (A.B.)
[2] National Interuniversity Consortium of Materials Science and Technology (INSTM), 50121 Florence, Italy
* Correspondence: matteo.savastano@unifi.it (M.S.); carla.bazzicalupi@unifi.it (C.B.)

Abstract: A tripodal anion receptor has been obtained by an easy and fast single-reaction synthesis from commercial reagents. The three ligand arms-bearing aromatic groups able to form anion–π interactions define ligand clefts where large anions, such as perchlorate and perrhenate, are included. We report here the synthesis of the ligand, its acid/base properties in an aqueous solution which has been used to direct the synthesis of anion complexes, and the crystal structure of the free ligand and its anion complexes $H_3L(ClO_4)_2 \cdot H_2O$ and $H_3L(ReO_4)_2$.

Keywords: anion complexes; perrhenate; perchlorate; anion–π; cleft system

1. Introduction

The design and synthesis of receptors for anionic species have attracted the interest of synthetic chemists since the dawn of supramolecular chemistry. Receptors of increasing complexity have been constructed, from simple linear or branched to more complex macrocyclic and macropolycyclic (cage-like) molecules, which have substantially contributed to forming the general concept that, with few exceptions, more structured and preorganized receptors form more stable anion complexes and provide more efficient molecular recognition processes. Indeed, macrocyclic and macropolycyclic molecules have proved to be very efficient anion receptors. Regrettably, they have the drawback that they often require lengthy and onerous synthetic procedures [1–11].

In this paper, we turn our attention toward a receptor that is easily accessible (from a synthetic point of view) and that, despite its open-chain structure, contains preorganized molecular clefts in which anions can be efficiently housed. This receptor (HL, Figure 1) is a tripodal molecule resembling the tetramine tren (tris(2-aminoethyl)amine), whose protonated species are known to form stable complexes with anions [12], but have a more rigid structure caused by the presence of three aromatic groups. HL can be easily prepared via a single-reaction functionalization of the commercial product N1,N1-bis(pyridin-2-ylmethyl)ethane-1,2-diamine with 6-amino-3,4-dihydro-3-methyl-5-nitroso-4-oxo-pyrimidine (Figure 1 and Section 2).

In an acidic solution, HL undergoes protonation forming ammonium groups able to bind anions via electrostatic attraction and the formation of salt bridges (hydrogen bonds between oppositely charged species). In addition, HL can also use its aromatic groups to bind anions via anion–π interactions. In particular, the pyrimidine inserted into the receptor was chosen precisely for this reason: being highly electron-poor, it has a marked tendency to form anion–π interactions, a characteristic confirmed by the crystal structures of several anion complexes with receptors containing this group [13–17]. Moreover, protonated pyridine groups have a non-negligible ability to form anion–π interactions [18–28].

In this paper, we perform a solid-state analysis of the anion-binding characteristics of H_3L^{2+}, the diprotonated form of HL, toward the monocharged tetrahedral ClO_4^- and ReO_4^- anions, observed in the crystal structures of $H_3L(ClO_4)_2 \cdot H_2O$ and $H_3L(ReO_4)_2$.

Citation: Savastano, M.; Cappanni, C.; Bazzicalupi, C.; Lofrumento, C.; Bianchi, A. Anion Coordination into Ligand Clefts. *Crystals* **2023**, *13*, 823. https://doi.org/10.3390/cryst13050823

Academic Editor: Ana M. Garcia-Deibe

Received: 31 March 2023
Revised: 3 May 2023
Accepted: 13 May 2023
Published: 16 May 2023

Copyright: © 2023 by the authors. Licensee MDPI, Basel, Switzerland. This article is an open access article distributed under the terms and conditions of the Creative Commons Attribution (CC BY) license (https://creativecommons.org/licenses/by/4.0/).

These anions were chosen because, in addition to their technological, biomedical, and environmental interest [29–37], they are challenging targets for binding as they have low charge density and poor hydrogen bonding ability [38,39]. Nevertheless, binding and even recognition of these anions was achieved in some cases, yet the receptors used were often more structurally complex and laborious to prepare [38–41].

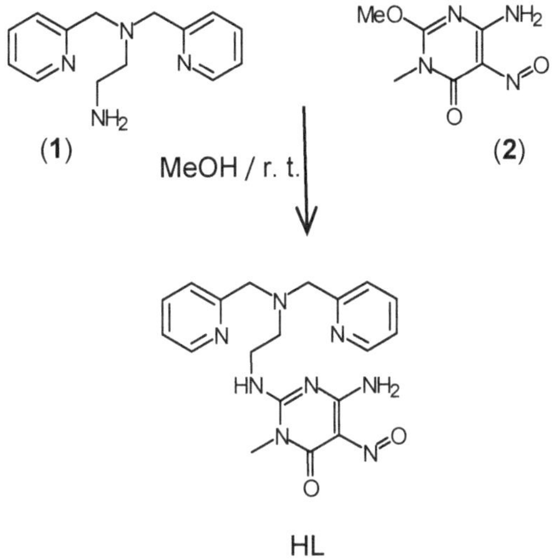

Figure 1. The receptor HL and the schematic procedure for its synthesis.

2. Materials and Methods

2.1. General

All starting materials were high-purity compounds purchased from commercial sources and were used without further purification. Compound 6-amino-3,4-dihydro-3-methyl-2-methoxy-5-nitroso-4-oxopyrimidine (2) was prepared according to a reported procedure [42]. Elemental analysis was performed with a Flash*Smart*™ Elemental Analyzer (Thermo Fisher Scientific, Monza, Italy).

2.2. Synthesis of HL (HL·EtOH·H_2O)

HL was synthesized as schematically shown in Figure 1. 0.58 g (3.1 mmol) of solid 2 was added in successive portions to a stirred solution of 1 (0.60 g, 2.5 mmol) in methanol (30 cm^3) at room temperature. After compound 2 was completely dissolved (about 90 min), 0.5 cm^3 of 37% NH_3 was added at room temperature to convert the excess of 2 into the insoluble 2,4-diamino-1-methyl-5-nitroso-6-oxopyrimidine derivative, and the resulting solution was left overnight at 4 °C. The suspension was then filtered and the solution was evaporated to dry under vacuum at room temperature to obtain HL as a deep purple solid compound that was recrystallized from ethanol. Yield 67%. ^{1}H NMR (D_2O, 400 MHz) δ 8.72 (d, 2H), 8.54 (t, 2H), 8.10 (d, 2H), 7.96 (t, 2H), 4.34 (s, 4H), 3.87 (t, 2H), 3.40 (s, 3H), 3.00 (t, 2H). Anal. calcd. for $C_{21}H_{30}N_8O_4$ (HL·EtOH·H_2O): C, 55.01; H, 6.59; N, 24.44. Found: C, 54.34; H, 6.52; N, 24.37. The crystals of this sample were suitable for X-ray.

2.3. Synthesis of $H_3L(ClO_4)_2·H_2O$

20 mg (0.044 mmol) of HL·EtOH·H_2O was dissolved in 5 cm^3 of water and the pH of the solution was brought to about 2.5 by the addition of diluted $HClO_4$. The crystals of $H_3L(ClO_4)_2·H_2O$ suitable for X-ray analysis were obtained by slow evaporation of this solution at room temperature. The crystals were filtered and air-dried. Yield 47%. Anal. calcd. for $C_{19}H_{26}N_8O_{11}Cl_2$: C, 37.21; H, 4.27; N, 18.27. Found: C, 37.09; H, 4.31; N, 18.16.

2.4. Synthesis of $H_3L(ReO_4)_2$

20 mg (0.044 mmol) of HL·EtOH·H$_2$O was dissolved in 5 cm^3 of water and the pH of the solution was brought to about 2.5 by the addition of diluted HCl. A two-fold excess of NaReO$_4$ was then added and the resulting solution was allowed to evaporate at room temperature to form the crystals of H$_3$L(ReO$_4$)$_2$ suitable for X-ray analysis. The crystals were filtered and air-dried. Yield 59%. Anal. calcd. for C$_{19}$H$_{24}$N$_8$O$_{10}$Re$_2$: C, 25.45; H, 2.70; N, 12.49. Found: C, 25.13; H, 2.78; N, 12.41.

2.5. Potentiometric Measurements

Ligand (HL) protonation constants were determined by means of potentiometric (pH-metric) titrations in 0.1 M NMe$_4$Cl aqueous solution at 298.1 ± 0.1 K using an automated apparatus and a procedure previously described [43]. The acquisition of the emf data was performed with the computer program PASAT [44,45]. The combined electrode (Metrohm 6.0262.100, Metrohm, Herisau, Switzerland) was calibrated as a hydrogen-ion concentration probe by titration of known amounts of HCl with CO$_2$-free NaOH solutions and by determining the equivalent point by Gran's method [46], which gives the standard potential, E°, and the ionic product of water (pK_w = 13.83(1) in 0.1 M NMe$_4$Cl at 298.1 K). The stability constants were calculated from the potentiometric data by means of the computer program HYPERQUAD [47]. The concentration of HL was about 1×10^{-3} M in all experiments. The studied pH range was 2.0–11.5. Three measurements were performed and used to determine the protonation constants. Titration curves and relative fittings are reported in Figure S1.

2.6. Spectroscopic Measurements

UV-vis absorption spectra were recorded at 298 K by using a Jasco V-670 spectrophotometer (Jasco Europe, Lecco, Italy). Ligand solution was [HL] = 2×10^{-5} M. Spectra were recorded in the pH range 0.80–7.24 (a) and 7.24–12.58 (b). FT-IR spectra were recorded at room temperature on crystalline samples in attenuated total reflectance (ATR) mode with an IRAffinity-1S instrument (Shimadzu, Milan, Italy).

2.7. Single-Crystal X-ray Diffraction Analyses

Purple crystals of HL·EtOH·H$_2$O and orange crystals of H$_3$L(ReO$_4$)$_2$ and H$_3$L(ClO$_4$)$_2$·H$_2$O were used for X-ray diffraction analysis. A summary of the crystallographic data is reported in Table 1. The integrated intensities were corrected for Lorentz and polarization effects and an empirical absorption correction was applied [48]. The structures were solved by direct methods (SHELXS-97) [49]. Refinements were performed by means of full-matrix least-squares using SHELXL Version 2014/7 [50]. All non-hydrogen atoms were anisotropically refined. Hydrogen atoms were usually introduced in a calculated position and their coordinates were refined according to the linked atoms. Two different conformations were found for the nitroso group in the perrhenate salt. Their population parameters were refined constraining their sum to 1 (0.538/0.462 at the end of refinement). The acidic hydrogen atoms and one water hydrogen in H$_3$L(ClO$_4$)$_2$·H$_2$O were localized in the Fourier difference maps, introduced in the calculation, and freely refined. In H$_3$L(ReO$_4$)$_2$, the acidic hydrogen atoms were introduced in the calculated position, linked to the pyridine nitrogens as found in the perchlorate salt, while in HL·EtOH·H$_2$O both water hydrogen atoms were not localized in the Fourier difference map and not introduced in the calculation. The labeling schemes for anion complexes are reported in Figure S2. Selected H-bond contacts are listed in Table S1. CCDC numbers 2253138-2253140.

Table 1. Crystal data and structure refinement for HL·EtOH·H$_2$O, H$_3$L(ReO$_4$)$_2$, and H$_3$L(ClO$_4$)$_2$·H$_2$O.

	HL·EtOH·H$_2$O	H$_3$L(ReO$_4$)$_2$	H$_3$L(ClO$_4$)$_2$·H$_2$O
Empirical formula	C$_{21}$H$_{30}$N$_8$O$_4$	C$_{19}$H$_{24}$N$_8$O$_{10}$Re$_2$	C$_{19}$H$_{26}$Cl$_2$N$_8$O$_{11}$
Formula weight	458.53	896.86	613.38
Temperature (K)	100(2)	100(2)	100(2)
space group	P-1	$P2_1/n$	$P2_1/n$
a (Å)	9.3960(4)	14.9381(9)	13.8500(5)
b (Å)	9.7782(4)	12.8194(6)	13.2210(5)
c (Å)	13.6308(6)	14.9659(7)	15.3154(6)
α (°)	76.058(2)	90	90
β (°)	70.650(2)	115.548(2)	113.661(2)
γ (°)	89.521(2)	90	90
Volume (Å^3)	1143.29(9)	2585.7(2)	2568.67(17)
Z	2	4	4
Independent reflections/R(int)	3500/0.1297	12,541/0.0662	4670/0.0583
μ (mm^{-1})	0.789/(Cu-kα)	9.423/(Mo-kα)	2.948/(Cu-kα)
R indices [I > 2σ(I)] *	R$_1$ = 0.0546 wR$_2$ = 0.1759	R$_1$ = 0.0639 wR$_2$ = 0.1575	R$_1$ = 0.0851 wR$_2$ = 0.2241
R indices (all data) *	R$_1$ = 0.1682 wR$_2$ = 0.2481	R$_1$ = 0.0963 wR$_2$ = 0.1802	R$_1$ = 0.0882 wR$_2$ = 0.2265

* R$_1$ = Σ ||Fo| − |Fc||/Σ |Fo|; wR$_2$ = [Σ w(Fo^2 − Fc^2)^2/Σ wFo^4]$^{\frac{1}{2}}$.

2.8. Hirshfeld Surface Analysis

The Hirshfeld surface and the fingerprint plots were calculated using the Crystalexplorer17 software [51].

3. Results and Discussion

3.1. Ligand Protonation

The acid/base properties of HL were determined in order to know in which pH regions the different protonated forms of the receptor are formed and, thus, to be able to select the appropriate conditions for the synthesis of the anion complexes.

The analysis of the potentiometric titration curves, performed with the program HYPERQUAD [47], showed that HL undergoes one deprotonation in alkaline solutions and three protonations in acidic solutions: the corresponding equilibrium constants are reported in Table 2.

Table 2. Protonation constants determined in 0.1 M NMe$_4$Cl aqueous solution at 298.1 ± 0.1 K. Standard deviations are reported in parentheses.

Equilibrium	LogK
L$^-$ + H$^+$ = HL	11.88(1)
HL + H$^+$ = H$_2$L$^+$	5.82(3)
H$_2$L$^+$ + H$^+$ = H$_3$L^{2+}	3.04(3)
H$_3$L^{2+} + H$^+$ = H$_4$L^{3+}	2.17(3)

In agreement with the behavior previously observed for other polyamines containing the same pyrimidine and pyridine groups [13,17,52–55], protonation occurring at high pH values (logK = 11.88, Table 2) is attributable to the secondary amine group, bound to the pyrimidine ring, which is deprotonated in very alkaline media, while protonation occurring in the lowest pH range (logK = 2.17, Table 2) is expected to involve the pyrimidine nitroso group (Scheme S1). The spectra of HL recorded in the near-UV at different pH values (Figure 2a,b) confirm this attribution of protonation sites. These spectra are characterized by three bands around 230, 257, and 330 nm which can be assigned to the allowed π–π* transitions between the π orbitals of the pyrimidine group and the overlap of the pyridine band at approximately 260 nm. As can be seen in Figure 2c, the pyrimidine bands undergo important variation with pH in acidic (pH < 3) and alkaline (pH > 11.5) solutions when protonation involves the pyrimidine chromophore while the spectra are almost invariant in the intermediate region. This intermediate region is dominated by the presence of the HL and its monoprotonated H_2L^+ species (Figure 2). As the pyridine band at 260 nm is expected to increase upon protonation [56], the invariance of this band suggests that protonation of HL to give H_2L^+ occurs on the tertiary amine group (Scheme S1). Further protonation to form H_3L^{2+} necessarily involves a pyridine group and, indeed, the 260 nm band increases but the increase is accentuated by the almost concomitant formation of H_4L^{3+} in which, as already commented, the pyrimidine chromophore becomes protonated on the nitroso group.

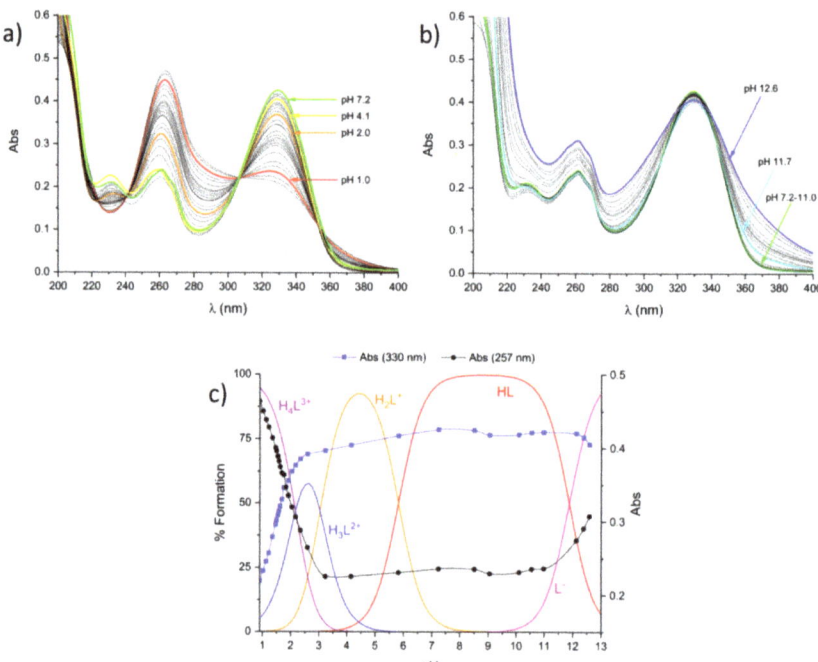

Figure 2. UV spectra of HL in the pH ranges 0.80–7.24 (**a**) and 7.24–12.58 (**b**). (**c**) Absorbances at 257 nm (●) and 330 nm (■) from the above spectra superimposed to the distribution diagram of the species formed by HL as a function of pH. [HL] = 2×10^{-5} M.

As for the location of the two ammonium groups in H_3L^{2+}, different alternatives should be considered: (i) one proton is located on the tertiary nitrogen atom (as in H_2L^+) and one on a pyridine; (ii) both pyridine groups are protonated; and (iii) protonation involves all three basic centers and there are several species in equilibrium with each other. The first alternative would be favored by the higher basicity of the tertiary amino group but could generate a strong repulsion between the neighboring ammonium groups, while, in the case of the second alternative, this electrostatic repulsion would be reduced since the two pyridines are the protonation sites which can stay at the maximum distance from each other. The third alternative would be a compromise between the first two. The crystal structures of perchlorate salt described below show that H_3L^{2+} contains two pyridinium groups, thus corroborating the second alternative (ii) (Scheme S1).

Based on the solution results, we decided to prepare the anion complexes at pH 2.5 in order to have in the solution the maximum abundance of the H_3L^{2+} species (Figure 2c). Although at such pH, H_2L^+ and H_4L^{3+} are also present in significant amounts, only the H$_3$L(ReO$_4$)$_2$ and H$_3$L(ClO$_4$)$_2$·H$_2$O salts were obtained and in fairly good yields (see Sections 2.3 and 2.4 above).

3.2. Crystallographic Study

3.2.1. Crystal Structure of HL·EtOH·H$_2$O

The crystal structure of the free ligand (HL), featuring co-crystallized ethanol and water molecules, is dominated by π–π stacking forces.

The arms of the tripodal ligand assume an overall Y conformation (Figure 3). One of the clefts of the ligand fully closes as one of the pyridine pendants gives face-to-face π–π stacking interaction with the nitroso–pyrimidine moiety (Figure 3, the dihedral angle between the rings' planes is 15.8 deg, the distance between ring centroids is 3.81 Å, and the shortest contact is C2···C11 3.389(4) Å), resulting in a local U fold. Above and below, the same aromatic groups are involved in pyrimidine–pyrimidine (ring centroids distance 3.46 Å) or pyridine–pyridine (ring centroids distance 3.89 Å) π-stacking contacts connecting centrosymmetric molecules which stack upon each other forming columns (Figure 4).

As required by both intrinsic ring polarity and consequent space group symmetry, both pyridine–pyridine and pyrimidine–pyrimidine contacts feature the two neighboring rings in antiparallel disposition. Antiparallel stacks are then connected by NH$_2$···N$_{ar}$ H-bonds (N2···N7' 2.909(5) Å).

The second pyridine arm is not involved in face-to-face π–π stacking contacts, but it rather protrudes from, and panels, said stacked columns, being, moreover, involved in hydrogen bonding with solvent molecules. Beyond the N8···O3' hydrogen bond (2.875(4) Å), neighboring hydrogen-bonded stacks interact with each other through further co-crystallized solvent-mediated (O3···O4 2.735(4) Å) H-bonds, namely N5···O3 and O4···N1' (2.791(3) Å and 2.963(4) Å, respectively (Figure S3, Table S1).

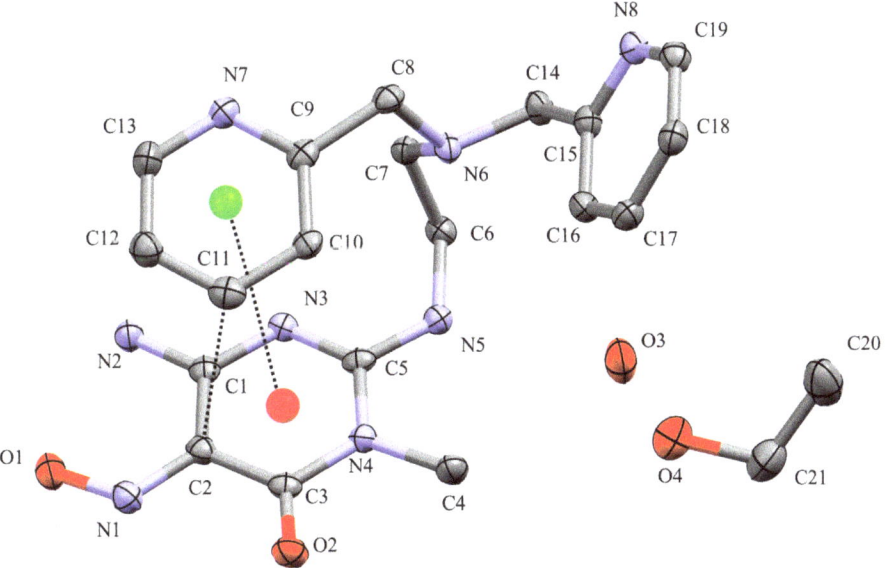

Figure 3. Asymmetric unit content of the HL·EtOH·H$_2$O crystal structure with atom labeling.

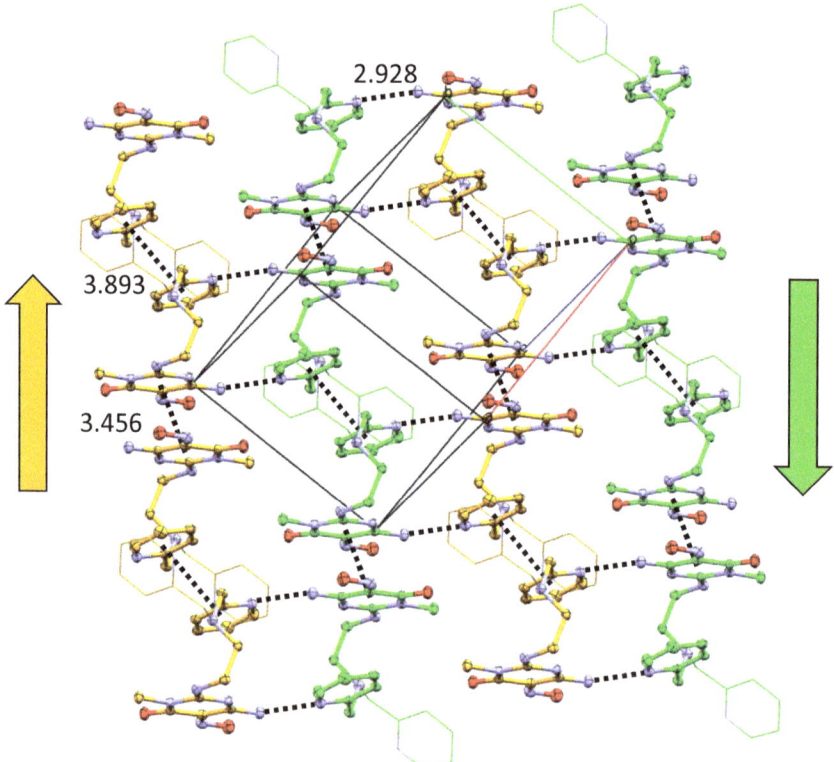

Figure 4. Repeats of antiparallel stacks of ligands (orange/green) featuring alternate head-to-tail pyridine/pyridine or pyrimidine/pyrimidine π–π stacking contacts in HL·EtOH·H$_2$O. H-bonded dimers alternate between inter- and intra-strand. The third arm of each ligand, not involved in strand formation, is shown as a wireframe. Hydrogen atoms and co-crystallized solvent molecules are omitted for clarity. Distances in Å.

3.2.2. Crystal Structures of the Anion Cleft Complexes H$_3$L(ReO$_4$)$_2$ and H$_3$L(ClO$_4$)$_2$·H$_2$O

The two structures can be considered isomorphous, having the same space group and almost the same cell parameters. The diprotonated ligand and the two anions (Figure 5) define very similar structural motifs: the U-shaped conformation seen for the not protonated ligand is now open and the three aromatic rings give two clamps gripping the two anions via anion–π interactions. The acidic hydrogens, which have been localized in the ΔF map for the perchlorate complex, are linked to both the pyridine nitrogen atoms, as expected from solution studies, and converge towards the nitroso oxygen of an adjacent symmetry-related ligand molecule (perchlorate complex: N7···O1 2.700(5) Å, N8···O1 2.793(6) Å; perrhenate complex: N7···O1 2.67(1) Å, N8···O1 2.62(1) Å, Table S1, see Figures S4 and S5 for the labeling scheme of both structures).

The pyrimidine–pyridine–pyridine ring sequence gives rise to the following dihedrals: 71.9/87.6 deg and 58.2/81.7 deg for the perchlorate and the perrhenate salt, respectively (Figure S6). Beyond these little differences, the ligand conformations are virtually identical (RMSD all atoms 0.2013), with their superposition, as shown in Figure 6, allowing us to discuss details of the anion-binding modes. The anion held by the pyridine–pyridine clamp has a very similar coordination mode, its central atom (Cl1 or Re2) being essentially located in the same position with respect to the ligand. These anions are equidistant from the two pyridine rings (Cl1 distance from pyridines centroids 4.089/4.108 Å, Re2 distance from pyridines centroids 4.172/3.958 Å) and give short contacts with the protonated nitrogen atoms and with the neighboring aromatic carbon (shortest contacts range from 3.007(5) Å to 3.146(6) Å in the perchlorate salt and from 2.90(1) Å to 3.18(1) Å in the perrhenate salt).

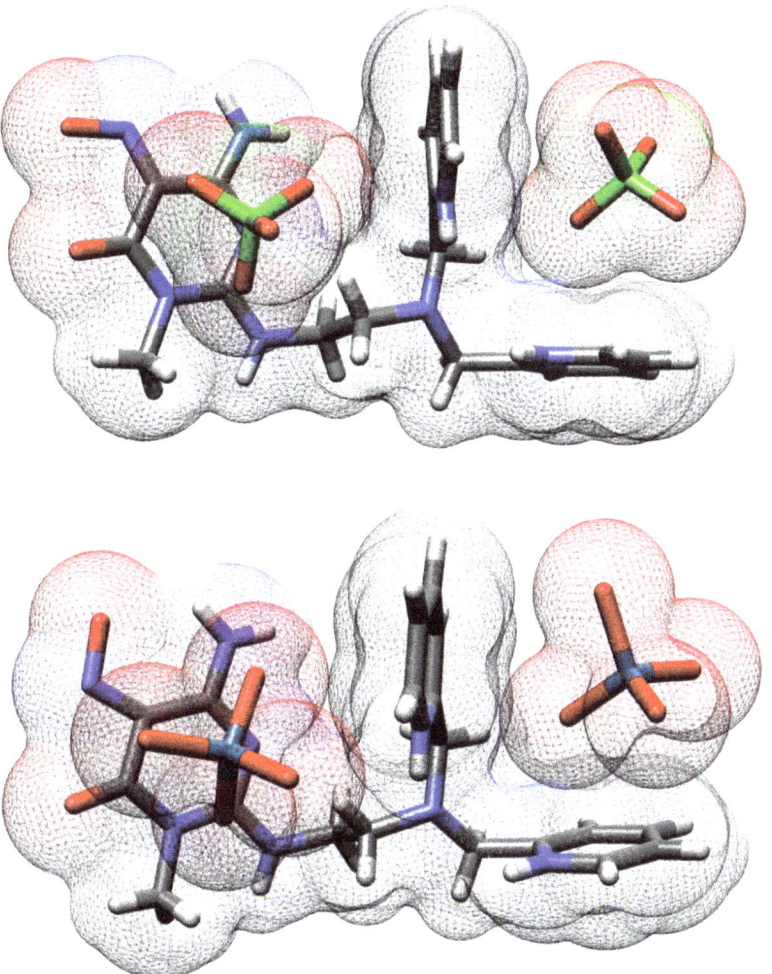

Figure 5. Synopsis of the perchlorate (**top**) and perrhenate (**bottom**) complexation inside the ligand clefts in the crystal structures of $H_3L(ClO_4)_2 \cdot H_2O$ and $H_3L(ReO_4)_2$.

The second cleft features two different aromatics, the polysubstituted pyrimidine being the most apt to anion–π binding. As a matter of fact, Cl2 perchlorate is closer to the pyrimidine ring (O24···N4 2.977(4) Å, O23···C2 3.290(6) Å) than to the pyridine ring (O22···N7 3.123(6) Å), as expected (Figure S4). Instead, the Re1 perrhenate anion is almost equally distant from both rings (O11···C13 3.18(1) Å, O13···N4 3.169(9) Å, Figure S5). The mutual positions of these anions with respect to the ligand cleft is arguably the major difference between the two crystal structures, the difference being manifest in Figure 6. Such tuning of the ligand–anion interactions seem to stem from two other supramolecular forces at work, hydrogen bonding, and anion–anion contacts among perrhenate species, as shown in Figure 7.

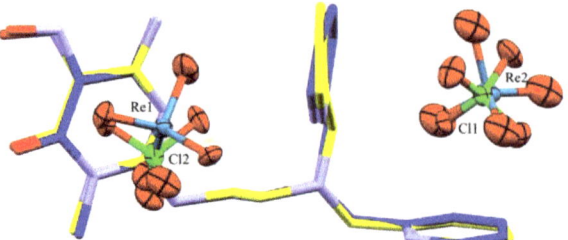

Figure 6. Superposition of the H_3L^{2+} cations found in the crystal structures of $H_3L(ReO_4)_2$ and $H_3L(ClO_4)_2 \cdot H_2O$ and differences in the relative positions of the anions inside the clefts.

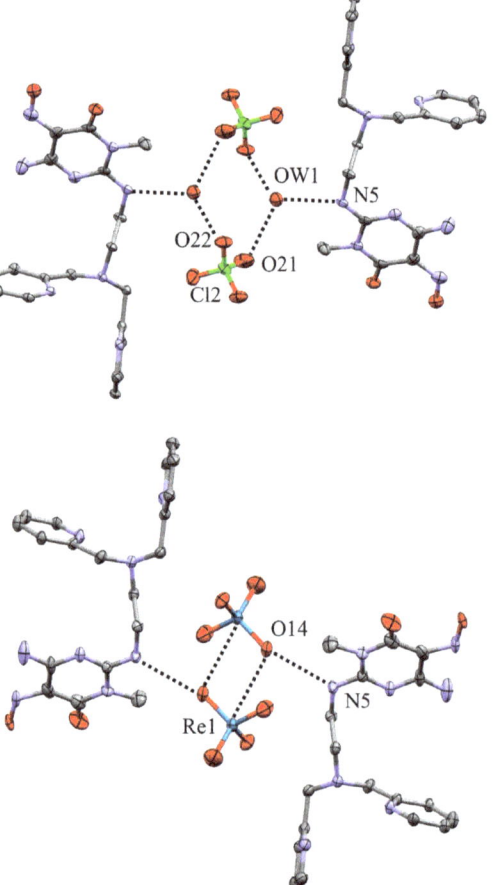

Figure 7. Details of further intermolecular contacts determining the exact disposition of anions bound within the pyrimidine–pyridine cleft in the crystal structures of $H_3L(ClO_4)_2 \cdot H_2O$ (above) and $H_3L(ReO_4)_2$ (below).

For both anions, the N5 nitrogen is involved in the binding. In the perrhenate case, this translates to a direct Re1–O14···HN5 hydrogen bond (O···N distance 3.044(8) Å, Table S1). The same interaction in the monohydrate perchlorate complex is mediated by the solvent molecule, which also bridges two symmetry-related anions (H-bond distances: N5···OW1 2.756(6) Å, Cl2–O21···OW1 2.831(6) Å, OW1···O22–Cl2 2.885(5) Å, Table S1). Perrhenate anions are also involved in weak mutual interactions with the establishment of a mutual O···Re/Re···O contact (O14···Re1′ 3.602(6) Å), which happens through one of the faces of the ReO_4^- tetrahedron. Similar contacts have been previously observed and recently dubbed matere bonds [57,58].

3.2.3. FT-IR Spectra Analysis

FT-IR spectra recorded on the crystalline samples of HL·EtOH·H_2O, $H_3L(ClO_4)_2$·H_2O, and $H_3L(ReO_4)_2$ are shown in Figure S7. If we exclude the signals attributable to ethanol (900, 1157, 1219, 1238 cm^{-1}), which are present in the spectrum of HL·EtOH·H_2O, and some alteration of the relative intensity of the bands, the three spectra are essentially very similar, the main differences consisting in the presence of the typical bands of ClO_4^- (3361, 1065, 613 cm^{-1}) and ReO_4^- (895 cm^{-1}) [59]. The intense bands at 1065 cm^{-1} and 895 cm^{-1} due to the characteristic Cl-O and Re-O asymmetric stretching vibrations of ClO_4^- and ReO_4^- in the spectra of $H_3L(ClO_4)_2$·H_2O and $H_3L(ReO_4)_2$, respectively, are red-shifted as compared to spectra reported for $MClO_4$ (1076–1100 cm^{-1}, M = Li$^+$, Na$^+$, NH_4^+ [60–62]) and $MReO_4$ (913–928 cm^{-1}, M = Na$^+$, K$^+$, NH_4^+ [63,64]), manifesting the effect of anion complexation within the protonated ligand clefts.

3.2.4. Hirshfeld Surface Analysis

The Hirshfeld surface of the free ligand and corresponding fingerprint plot are reported in Figure 8.

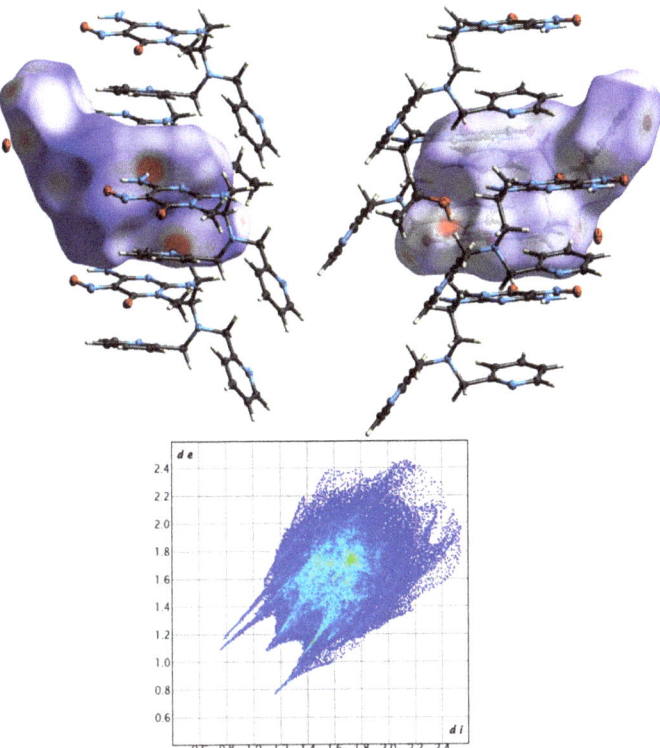

Figure 8. (**Top**): views of the HL Hirshfeld surface (d_{norm} coloring) in the HL·EtOH·H_2O crystal structure. (**Bottom**): global fingerprint plot.

The Hirshfeld surface of HL is a rather flat block, delimited on top and bottom sides by intracolumn π–π stacking interactions and, on its sides, by intercolumn interactions, the most significant of which are the above-mentioned N2···N7 hydrogen bonds.

The fingerprint plot shows both typical H-bond tips and a manifest π–π stacking area (green portion of the graph centered at {1.7,1.7} coordinates). These are the same interaction types emerging from overall contact percentages (Table S2), with the further addition of H···O contacts due to hydrogen bonding to solvent molecules. The pseudosymmetry of the fingerprint graph is only slightly broken by the presence of the co-crystallized solvent molecules.

The Hirshfeld surface of the diprotonated ligand in the perchlorate and perrhenate complexes and corresponding fingerprint plots are shown in Figures 9 and 10, respectively.

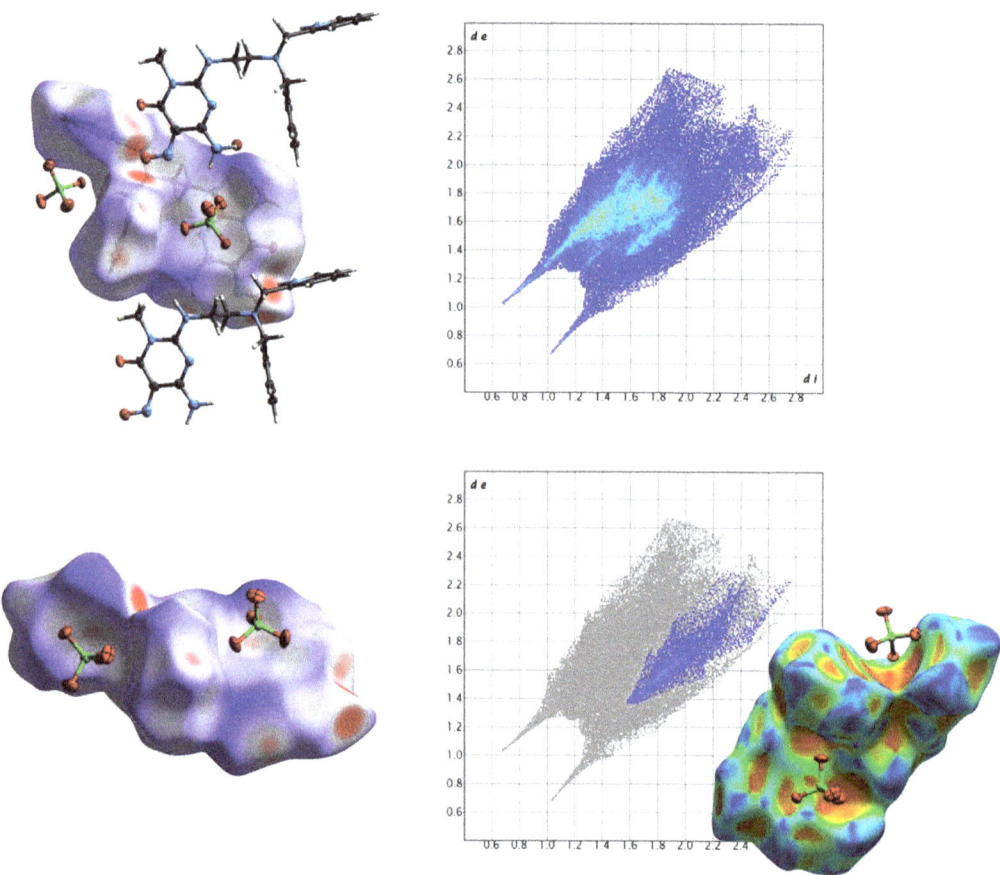

Figure 9. (**Top**): views of the H_3L^{2+} Hirshfeld surface (d_{norm} coloring) in the $H_3L(ClO_4)_2 \cdot H_2O$ crystal structure with main contacts highlighted and the full fingerprint plot. (**Bottom**): focus on the anion binding inside the clefts with detail of the C···O contact portion of the fingerprint plot. A shape index coloring scheme is also used to highlight the anion inclusion inside the binding pockets.

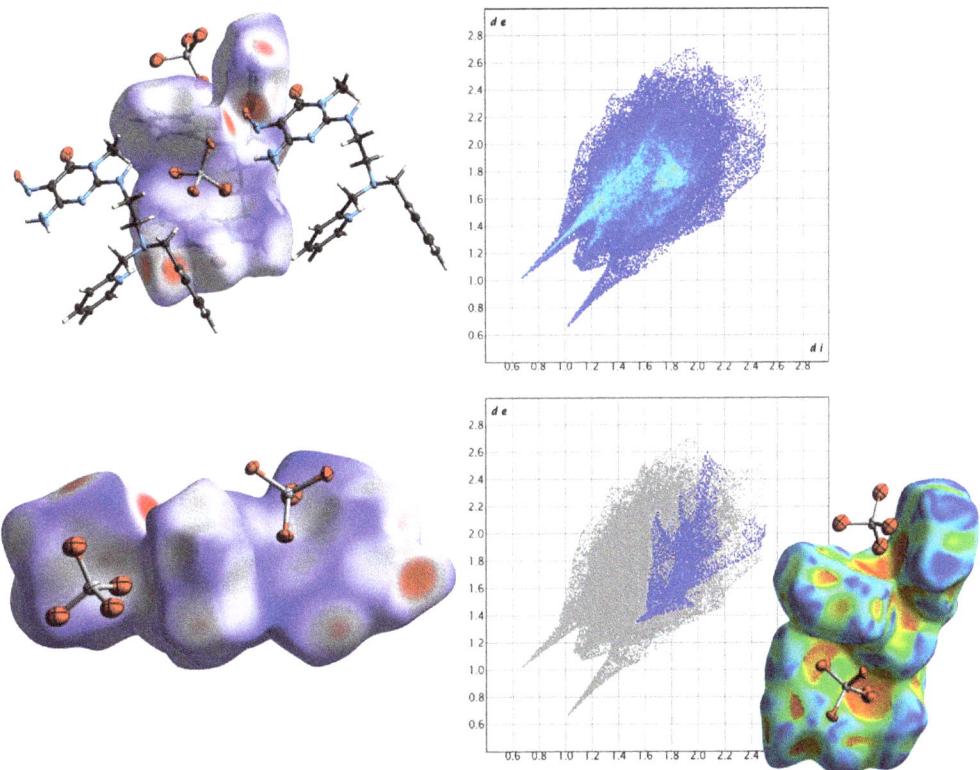

Figure 10. (**Top**): views of the H_3L^{2+} Hirshfeld surface (d_{norm} coloring) in the $H_3L(ReO_4)_2$ crystal structure with main contacts highlighted and the full fingerprint plot. (**Bottom**): focus on the anion binding inside the clefts with detail of the C···O contact portion of the fingerprint plot. A shape index coloring scheme is also used to highlight the anion inclusion inside the binding pockets.

Main interaction hotspots, as judged from the d_{norm} parameter, belong to the ligand–ligand H-bonds, namely among pyridinium sites and NO oxygen.

Anions are found in molecular clefts. The anion–π interaction is not as immediately manifest due to the simultaneous presence of H-bond type interactions, which, being both stronger and more directional, are easier to spot both from d_{norm} hotspots on the Hirshfeld surface and as sharp tips in the fingerprint plots. However, if C···O, i.e., anion–π type contacts, are highlighted, the interaction is observed to be significantly represented in the fingerprint plot [22], and to account for a significant portion of the global Hirshfeld surface area (5.9% and 6.3% for ClO_4^- and ReO_4^-, respectively, Tables S3 and S4). The supramolecular hosting of anionic species is here perhaps better visualized through the shape index coloring, which is effective at showing inward/outward local bending of the Hirshfeld surface itself. As can be observed, anion-binding clefts are among the areas that show the most significant inward bending (red), signaling how the ligand's surface engulfs the anions (Figures 9 and 10 bottom).

Of course, the contribution of these weak forces combines and synergizes with the electrostatic attraction generated between anions and ligands as a consequence of the positive charge introduced by protonation on the pyridine groups, which reverberates throughout the ring as evidenced by the calculated potential electrostatic surfaces (ESP) shown in Figure S8.

4. Conclusions

Tripodal ligands are convenient building blocks for easy and fast synthesis of efficient anion receptors. An example is given here by the HL ligand, characterized by a tren-like (tren = tris(2-aminoethyl)amine) structure bearing two pyridine and one pyrimidine terminal functionalities, whose synthesis was accomplished by means of a single reaction from commercial products. The crystal structure of the free ligand shows that the three arms decorated with aromatic groups define molecular clefts where host species can be included. Upon protonation, the ligand opens these clefts as wide as necessary for tight inclusion of large anions such as perchlorate and perrhenate, as shown in the solid state by the crystal structures of $H_3L(ClO_4)_2 \cdot H_2O$ and $H_3L(ReO_4)_2$. The ligand arms act as clamps gripping the two anions via anion–π interactions which are the main forces, in addition to the electrostatic attraction exerted by the ammonium groups that participate in strengthening the host–guest association. Intermolecular hydrogen bonding, π-stacking, and anion–anion interactions, in the case of perrhenate, contribute to stabilizing the crystal packing.

Supplementary Materials: The following supporting information can be downloaded at: https://www.mdpi.com/article/10.3390/cryst13050823/s1, Figure S1: Titration curves. Figure S2: Labelling schemes for anion complex structures. Figure S3: Pairs of hydrogen-bonded ligands (same color) joined in an H-bond network via co-crystallized solvent molecules. Figure S4: Selected short contacts (H-bonds and anion–π) in the perchlorate complex. Figure S5: Selected short contacts (H-bonds and anion–π) in the perrhenate complex. Figure S6: Details of the dihedral angles formed by the pyridine–pyridine–pyrimidine rings' sequence in the perchlorate and perrhenate salts. Figure S7: FT-IR spectra of crystalline $HL \cdot EtOH \cdot H_2O$, $H_3L(ClO_4)_2 \cdot H_2O$ and $H_3L(ReO_4)_2$. Figure S8: Electrostatic potential surfaces of HL and H_3L^{2+}. Scheme S1: Ligand protonation scheme. Table S1: Selected H-bond contacts. Table S2: Atoms in contact with the HL Hirshfeld surface in the $HL \cdot EtOH \cdot H_2O$ crystal structure. Table S3: Atoms in contact with the H_3L Hirshfeld surface in the $H_3L(ClO_4)_2 \cdot H_2O$ crystal structure. Table S4: Atoms in contact with the H_3L Hirshfeld surface in the $H_3L(ReO_4)_2$ crystal structure. Cif files of the three structures.

Author Contributions: Conceptualization, M.S., C.B. and A.B.; methodology, M.S., C.L. and C.B.; validation, M.S., C.B. and A.B.; investigation, C.C., C.L. and M.S.; data curation, C.C., C.L., M.S. and C.B.; writing—review and editing, M.S., C.B. and A.B.; funding acquisition, A.B. All authors have read and agreed to the published version of the manuscript.

Funding: This research received no external funding.

Data Availability Statement: Not applicable.

Acknowledgments: The financial support provided by the MUR—Dipartimenti di Eccellenza 2023–2027 (DICUS 2.0) to the Department of Chemistry "Ugo Schiff" of the University of Florence is acknowledged. This contribution is also based upon work from COST Action CA18202, NECTAR—Network for Equilibria and Chemical Thermodynamics Advanced Research, supported by COST (European Cooperation in Science and Technology).

Conflicts of Interest: The authors declare no conflict of interest.

References

1. Bianchi, A.; Bowman-James, K.; Garcia-España, E. (Eds.) *Supramolecular Chemistry of Anions*; Wiley-VCH: New York, NY, USA, 1997.
2. Sessler, J.L.; Gale, P.A.; Cho, W.S. *Anion Receptor Chemistry (Monographs in Supramolecular Chemistry)*; RSC Publishing: Cambridge, UK, 2006.
3. Bowman-James, K.; Bianchi, A.; Garcia-España, E. (Eds.) *Anion Coordination Chemistry*; Wiley-VCH: New York, NY, USA, 2012.
4. Zhao, J.; Yang, D.; Yang, X.-J.; Wu, B. Anion coordination chemistry: From recognition to supramolecular assembly. *Coord. Chem. Rev.* **2019**, *378*, 415–444. [CrossRef]
5. He, Q.; Tu, P.; Sessler, J.L. Supramolecular Chemistry of Anionic Dimers, Trimers, Tetramers, and Clusters. *Chem* **2018**, *4*, 46–93. [CrossRef] [PubMed]
6. Custelcean, R. Anion encapsulation and dynamics in self-assembled coordination cages. *Chem. Soc. Rev.* **2014**, *43*, 1813–1824. [CrossRef] [PubMed]

7. Bowman-James, K. Alfred Werner Revisited: The Coordination Chemistry of Anions. *Acc. Chem. Res.* **2005**, *38*, 671–678. [CrossRef]
8. Ilioudis, C.A.; Steed, J.W. Organic macrocyclic polyamine-based receptors for anions. *J. Supramol. Chem.* **2001**, *1*, 165–187. [CrossRef]
9. Llinares, J.M.; Powell, D.; Bowman-James, K. Ammonium based anion receptors. *Coord. Chem. Rev.* **2003**, *240*, 57–75. [CrossRef]
10. Gale, P.A. Anion receptor chemistry: Highlights from 1999. *Coord. Chem. Rev.* **2001**, *213*, 79–128. [CrossRef]
11. Dietrich, B. Design of anion receptors: Applications. *Pure Appl. Chem.* **1993**, *65*, 1457–1464. [CrossRef]
12. Bazzicalupi, C.; Bencini, A.; Bianchi, A.; Danesi, A.; Giorgi, C.; Valtancoli, B. Anion Binding by Protonated Forms of the Tripodal Ligand Tren. *Inorg. Chem.* **2009**, *48*, 2391–2398. [CrossRef]
13. García-Martín, J.; López-Garzón, R.; Godino-Salido, M.L.; Cuesta, R.; Gutiérrez-Valero, M.D.; Arranz-Mascarós, P.; Stoeckli-Evans, H. Adsorption of Zn^{2+} and Cd^{2+} from Aqueous Solution onto a Carbon Sorbent Containing a Pyrimidine–Polyamine Conjugate as Ion Receptor. *Eur. J. Inorg. Chem.* **2005**, *2005*, 3093–3103. [CrossRef]
14. Arranz, P.; Bianchi, A.; Cuesta, R.; Giorgi, C.; Godino, M.L.; Gutiérrez, M.D.; López, R.; Santiago, A. Binding and Removal of Sulfate, Phosphate, Arsenate, Tetrachloromercurate, and Chromate in Aqueous Solution by Means of an Activated Carbon Functionalized with a Pyrimidine-Based Anion Receptor (HL). Crystal Structures of $[H_3L(HgCl_4)]_3 \cdot H_2O$ and $[H_3L(HgBr_4)]_3 \cdot H_2O$ Showing Anion-π Interactions. *Inorg. Chem.* **2010**, *49*, 9321–9332.
15. Arranz, P.; Bazzicalupi, C.; Bianchi, A.; Giorgi, C.; Godino, M.L.; Gutiérrez, M.D.; López, R.; Savastano, M. Thermodynamics of Anion−π Interactions in Aqueous Solution. *J. Am. Chem. Soc.* **2013**, *135*, 102–105. [CrossRef] [PubMed]
16. Savastano, M.; Arranz, P.; Bazzicalupi, C.; Bianchi, A.; Giorgi, C.; Godino, M.L.; Gutiérrez, M.D.; López, R. Binding and removal of octahedral, tetrahedral, square planar and linear anions in water by means of activated carbon functionalized with a pyrimidine-based anion receptor. *RSC Adv.* **2014**, *4*, 58505–58513. [CrossRef]
17. Savastano, M.; Arranz-Mascarós, P.; Bazzicalupi, C.; Clares, M.P.; Godino-Salido, M.L.; Gutiérrez-Valero, M.D.; Guijarro, L.; Bianchi, A.; Garcia-España, E.; López-Garzón, R. Polyfunctional Tetraaza-Macrocyclic Ligands: Zn(II), Cu(II) Binding and Formation of Hybrid Materials with Multiwalled Carbon Nanotubes. *ACS Omega* **2017**, *2*, 3868–3877. [CrossRef] [PubMed]
18. Kou, X.; Ma, Y.; Pan, C.; Huang, Y.; Duan, Y.; Yang, Y. Effects of the Cationic Structure on the Adsorption Performance of Ionic Polymers toward Au(III): An Experimental and DFT Study. *Langmuir* **2022**, *38*, 6116–6127. [CrossRef] [PubMed]
19. Xu, Y.; Dang, D.; Zhang, N.; Zhang, J.; Xu, R.; Wang, Z.; Zhou, Y.; Zhang, H.; Liu, H.; Yang, Z.; et al. Aggregation-Induced Emission (AIE) in Superresolution Imaging: Cationic AIE Luminogens (AIEgens) for Tunable Organelle-Specific Imaging and Dynamic Tracking in Nanometer Scale. *ACS Nano* **2022**, *16*, 5932–5942. [CrossRef] [PubMed]
20. Das, A.; Sharma, P.; Frontera, A.; Verma, A.K.; Barcelo-Oliver, M.; Hussain, S.; Bhattacharyya, M.K. Energetically significant nitrile···nitrile and unconventional C–H···π(nitrile) interactions in pyridine based Ni(II) and Zn(II) coordination compounds: Antiproliferative evaluation and theoretical studies. *J. Mol. Struct.* **2021**, *1223*, 129246. [CrossRef]
21. Qi, B.; An, S.; Luo, J.; Liu, T.; Song, Y.-F. Enhanced Macroanion Recognition of Superchaotropic Keggin Clusters Achieved by Synergy of Anion–π and Anion–Cation Interactions. *Chem. Eur. J.* **2020**, *26*, 16802–16810. [CrossRef]
22. Martínez-Camarena, Á.; Savastano, M.; Bazzicalupi, C.; Bianchi, A.; García-España, E. Stabilisation of Exotic Tribromide (Br_3^-) Anions via Supramolecular Interaction with a Tosylated Macrocyclic Pyridinophane. A Serendipitous Case. *Molecules* **2020**, *25*, 3155. [CrossRef]
23. Savastano, M.; Bazzicalupi, C.; Gellini, C.; Bianchi, A. Infinite supramolecular pseudo-polyrotaxane with poly[3]catenane axle: Assembling nanosized rings from mono- and diatomic I^- and I_2 tectons. *Chem. Commun.* **2020**, *56*, 551–554. [CrossRef]
24. Savastano, M.; Bazzicalupi, C.; Gellini, C.; Bianchi, A. Genesis of complex polyiodide networks: Insights on the blue box/I^-/I_2 ternary system. *Crystals* **2020**, *10*, 387. [CrossRef]
25. Savastano, M.; Bazzicalupi, C.; Bianchi, A. Porous frameworks based on supramolecular ball joints: Bringing flexibility to ordered 3D lattices. *Chem. Eur. J.* **2020**, *26*, 5994–6005. [CrossRef] [PubMed]
26. Giese, M.; Albrecht, M.; Repenko, T.; Sackmann, J.; Valkonen, A.; Rissanen, K. Single-Crystal X-ray Diffraction and Solution Studies of Anion–π Interactions in N-(Pentafluorobenzyl)pyridinium Salts. *Eur. J. Org. Chem.* **2014**, *2014*, 2435–2442. [CrossRef]
27. Biswas, C.; Drew, M.G.B.; Escudero, D.; Frontera, A.; Ghosh, A. Anion–π, Lone-Pair–π, π–π and Hydrogen-Bonding Interactions in a Cu^{II} Complex of 2-Picolinate and Protonated 4,4′-Bipyridine: Crystal Structure and Theoretical Studies. *Eur. J. Inorg. Chem.* **2009**, *2009*, 2238–2246. [CrossRef]
28. de Hoog, P.; Gamez, P.; Mutikainen, I.; Turpeinen, U.; Reedijk, J. An Aromatic Anion Receptor: Anion–p Interactions do Exist. *Angew. Chem. Int. Ed.* **2004**, *43*, 5815–5817. [CrossRef]
29. Krishnan Suresh, K.; Subbiah, K.; Kalivel, P.; Ayyasamy, S.; Palanivel, S. Environmental occurrence, toxicity and remediation of perchlorate—A review. *Chemosphere* **2023**, *311*, 137017.
30. Shukla, M.K.; Boddu, V.M.; Steevens, J.A.; Damavarapu, R.; Leszczynski, J. (Eds.) Energetic Materials: From Cradle to Grave. In *Challenges and Advances in Computational Chemistry and Physics*; Springer: Berlin/Heidelberg, Germany, 2018; Volume 25.
31. Vogt, H.; Balej, J.; Bennett, J.E.; Wintzer, P.; Sheikh, S.A.; Gallone, P. Chlorine Oxides and Chlorine Oxygen Acids. In *Ullmann's Encyclopedia of Industrial Chemistry*; Wiley-VCH: New York, NY, USA, 2002.
32. Gu, B.; Coates, J.D. (Eds.) *Perchlorate. Environmental Occurrence, Interactions and Treatment*; Springer Nature: Berlin/Heidelberg, Germany, 2006.

33. Da, H.-J.; Yang, C.-X.; Yan, X.-P. Cationic Covalent Organic Nanosheets for Rapid and Selective Capture of Perrhenate: An Analogue of Radioactive Pertechnetate from Aqueous Solution. *Environ. Sci. Technol.* **2019**, *53*, 5212–5220. [CrossRef]
34. Kohlickova, M.; Jedinakova-Krizova, V.; Melichar, F. Rhenium complexes in nuclear medicine. *Chem. Listy* **2000**, *94*, 151–158.
35. Volkert, W.A.; Hoffman, T.J. Therapeutic Radiopharmaceuticals. *Chem. Rev.* **1999**, *99*, 2269–2292. [CrossRef]
36. Colton, R. *The Chemistry of Rhenium and Technetium*, 1st ed.; John Wiley and Sons Ltd.: Hoboken, NJ, USA; Interscience Publishers: New York, NY, USA, 1965.
37. Banerjee, D.; Kim, D.; Schweiger, M.J.; Krugerc, A.A.; Thallapally, P.K. Removal of TcO_4^- ions from solution: Materials and future outlook. *Chem. Soc. Rev.* **2016**, *45*, 2724–2739. [CrossRef]
38. Katayev, E.A.; Kolesnikov, G.V.; Sessler, J.L. Molecular recognition of pertechnetate and perrhenate. *Chem. Soc. Rev.* **2009**, *38*, 1572–1586. [CrossRef] [PubMed]
39. Gale, P.A.; Howe, N.W.; Wu, X. Anion Receptor Chemistry. *Chem* **2016**, *1*, 351–422. [CrossRef]
40. Zhou, W.; Li, A.; Gale, P.A.; He, Q. A highly selective superphane for ReO_4^- recognition and extraction. *Cell Rep. Phys. Sci.* **2022**, *3*, 100875. [CrossRef]
41. Macreadie, L.K.; Gilchrist, A.M.; McNaughton, D.A.; Ryder, W.G.; Fares, M.; Gale, P.A. Progress in anion receptor chemistry. *Chem* **2022**, *8*, 46–118. [CrossRef]
42. Low, N.J.; López, M.D.; Arranz, P.; Cobo, J.; Godino, M.L.; López, R.; Gutiérrez, M.D.; Melguizo, M.; Ferguson, G.; Glidewell, C. N-(6-Amino-3,4-dihydro-3-methyl-5-nitroso-4-oxopyrimidin-2-yl) derivatives of glycine, valine, serine, threonine and methionine: Interplay of molecular, molecular-electronic and supramolecular structures. *Acta Cryst. B* **2000**, *56*, 882–892. [CrossRef]
43. Bazzicalupi, C.; Bianchi, A.; Biver, T.; Giorgi, C.; Santarelli, S.; Savastano, M. Formation of double-strand dimetallic helicates with a terpyridine-based macrocycle. *Inorg. Chem.* **2014**, *53*, 12215–12224. [CrossRef]
44. Fontanelli, M.; Micheloni, M. *Proceedings of the I Spanish-Italian Congress on Thermodynamics of Metal Complexes*; Diputación de Castellón: Castellón, Spain, 1990; pp. 41–43.
45. Savastano, M.; Fiaschi, M.; Ferraro, G.; Gratteri, P.; Mariani, P.; Bianchi, A.; Bazzicalupi, C. Sensing Zn^{2+} in Aqueous Solution with a Fluorescent Scorpiand Macrocyclic Ligand Decorated with an Anthracene Bearing Tail. *Molecules* **2020**, *25*, 1355. [CrossRef]
46. Gran, G. Determination of the equivalence point in potentiometric titrations. Part II. *Analyst* **1952**, *77*, 661–671. [CrossRef]
47. Gans, P.; Sabatini, A.; Vacca, A. Investigation of equilibria in solution. Determination of equilibrium constants with the HYPERQUAD suite of programs. *Talanta* **1996**, *43*, 1739–1753. [CrossRef]
48. Krause, L.; Herbst-Irmer, R.; Sheldrick, G.M.; Stalke, D. Comparison of silver and molybdenum microfocus X-ray sources for single-crystal structure determination. *J. Appl. Cryst.* **2015**, *48*, 3–10. [CrossRef]
49. Sheldrick, G.M. A short history of SHELX. *Acta Crystallogr. A* **2008**, *64*, 112–122. [CrossRef] [PubMed]
50. Sheldrick, G.M. Crystal structure refinement with SHELXL. *Acta Cryst.* **2015**, *C71*, 3–8.
51. Turner, M.J.; McKinnon, J.J.; Wolff, S.K.; Grimwood, D.J.; Spackman, P.R.; Jayatilaka, D.; Spackman, M.A. *CrystalExplorer17*; University of Western Australia: Crawley, WA, Australia, 2017; Available online: https://crystalexplorer.net (accessed on 6 February 2023).
52. Savastano, M.; Arranz-Mascarós, P.; Clares, M.P.; Cuesta, R.; Godino-Salido, M.L.; Guijarro, L.; Gutiérrez-Valero, M.D.; Inclán, M.; Bianchi, A.; García-España, E.; et al. A new heterogeneous catalyst obtained via supramolecular decoration of graphene with a Pd^{2+} azamacrocyclic complex. *Molecules* **2019**, *24*, 2714. [CrossRef] [PubMed]
53. Savastano, M.; Arranz-Mascarós, P.; Bazzicalupi, C.; Clares, M.P.; Godino-Salido, M.L.; Gutiérrez-Valero, M.D.; Inclán, M.; Bianchi, A.; Garcia-España, E.; López-Garzón, R. Construction of green nanostructured heterogeneous catalysts via non-covalent surface decoration of multi-walled carbon nanotubes with Pd(II) complexes of azamacrocycles. *J. Catal.* **2017**, *353*, 239–249. [CrossRef]
54. Godino-Salido, M.-L.; Santiago-Medina, A.; Arranz-Mascarós, P.; López-Garzón, R.; Gutiérrez-Valero, M.D.; Melguizo, M.; López-Garzón, F.J. Novel active carbon/crown ether derivative hybrid material for the selective removal of Cu(II) ions: The crucial role of the surface chemical functions. *Chem. Eng. Sci.* **2014**, *114*, 94–104. [CrossRef]
55. Savastano, M.; Passaponti, M.; Giurlani, W.; Lari, L.; Bianchi, A.; Innocenti, M. Multi-Walled Carbon Nanotubes Supported Pd(II) Complexes: A Supramolecular Approach towards Single-Ion Oxygen Reduction Reaction Catalysts. *Energies* **2020**, *13*, 5539. [CrossRef]
56. Hughes, E.B.; Jellinek, H.H.G.; Ambrose, B.A. Pyridine. Ultraviolet Absorption Spectrum and Dissociation Constant. *J. Phys. Chem.* **1949**, *53*, 410–414. [CrossRef]
57. Daolio, A.; Pizzi, A.; Terraneo, G.; Frontera, A.; Resnati, G. Anion···Anion Interaction involving σ-Holes of Perrhenate, Pertechnetate and Permanganate Anions. *ChemPhysChem* **2021**, *22*, 2281–2285. [CrossRef]
58. Gomila, R.M.; Frontera, A. Matere Bonds vs. Multivalent Halogen and Chalcogen Bonds: Three Case Studies. *Molecules* **2022**, *27*, 6597. [CrossRef]
59. Nyquist, R.A.; Kagel, R.O. *Infrared and Raman Spectra of Inorganic Compounds and Organic Salts*, 2nd ed.; Academic Press: New York, NY, USA, 1971.
60. Sima, L.H.; Ganb, S.N.; Chana, C.H.; Yahyab, R. ATR-FTIR studies on ion interaction of lithium perchlorate in polyacrylate/poly(ethylene oxide) blends. *Spectrochim. Acta Part A* **2010**, *76*, 287–292. [CrossRef]
61. Chandrawanshi, S.; Verma, S.K.; Deb, M.K. Collective Ion-Pair Single-Drop Microextraction Attenuated Total Reflectance Fourier Transform Infrared Spectroscopic Determination of Perchlorate in Bioenvironmental Samples. *J. AOAC Int.* **2018**, *101*, 1145–1155. [CrossRef] [PubMed]

62. Hori, K.; Iwama, A.; Fukuda, T. FTIR Spectroscopic Study on the Interaction between Ammonium Perchlorate and Bonding Agents. *Propell. Explos. Pyrot.* **1990**, *15*, 99–102. [CrossRef]
63. Gonzalez-Rodriguez, J.; Pepper, K.; Baron, M.G.; Mamo, S.K.; Simons, A.M. Production and Analysis of Recycled Ammonium Perrhenate from CMSX-4 superalloys. *Open Chem.* **2018**, *16*, 1298–1306. [CrossRef]
64. Gassman, P.-L.; McCloy, J.S.; Soderquist, C.Z.; Schweiger, M.J. Raman analysis of perrhenate and pertechnetate in alkali salts and borosilicate glasses. *J. Raman Spectrosc.* **2014**, *45*, 139–147. [CrossRef]

Disclaimer/Publisher's Note: The statements, opinions and data contained in all publications are solely those of the individual author(s) and contributor(s) and not of MDPI and/or the editor(s). MDPI and/or the editor(s) disclaim responsibility for any injury to people or property resulting from any ideas, methods, instructions or products referred to in the content.

Article

Structural and Magnetic Properties of the {Cr(pybd)$_3$[Cu(cyclen)]$_2$}(BF$_4$)$_4$ Heteronuclear Complex

Fabio Santanni [1,2,*], Laura Chelazzi [1], Lorenzo Sorace [1,2], Grigore A. Timco [3] and Roberta Sessoli [1,2,*]

1. Dipartimento di Chimica "Ugo Schiff"—DICUS, Università degli Studi di Firenze, Via della Lastruccia 3-13, 50019 Firenze, Italy; laura.chelazzi@unifi.it (L.C.); lorenzo.sorace@unifi.it (L.S.)
2. Consorzio Interuniversitario Nazionale di Scienza e Tecnologia dei Materiali—INSTM, Via G. Giusti 9, 50121 Firenze, Italy
3. Department of Chemistry, Photon Science Institute, The University of Manchester, Oxford Road, Manchester M13 9PL, UK; grigore.timco@manchester.ac.uk
* Correspondence: fabio.santanni@unifi.it (F.S.); roberta.sessoli@unifi.it (R.S.)

Abstract: Heterotopic ligands containing chemically different binding centers are appealing candidates for obtaining heteronuclear metal complexes. By exploiting this strategy, it is possible to introduce different paramagnetic centers characterized by specific anisotropic magnetic properties that make them distinguishable when weakly magnetically coupled. This molecular approach has great potential to yield multi-spin adducts capable of mimicking logical architectures necessary for quantum information processing (QIP), i.e., quantum logic gates. A possible route for including a single-ion magnetic center within a finite-sized heterometallic compound uses the asymmetric (1-pyridyl)-butane-1,3-dione (pybd) ligand reported in the literature for obtaining Cr^{3+}–Cu^{2+} metallocages. To avoid the formation of cages, we adopted the cyclen (1,4,7,10-tetraazacyclododecane) ligand as a "capping" agent for the Cu^{2+} ions. We report here the structural and magnetic characterization of the unprecedented adduct {Cr(pybd)$_3$[Cu(cyclen)]$_2$}(BF$_4$)$_4$, whose structure is characterized by a central Cr^{3+} ion in a distorted octahedral coordination environment and two peripheral Cu^{2+} ions with square-pyramidal coordination geometries. As highlighted by Continuous Wave Electron Paramagnetic Resonance (EPR) spectroscopy and Direct Current (DC) magnetometry measurements, this adduct shows negligible intramolecular magnetic couplings, and it maintains the characteristic EPR signals of Cr^{3+} and Cu^{2+} moieties when diluted in frozen solutions.

Keywords: heterometallic Cr-Cu complexes; powder X-ray diffraction structural determination; magnetism; EPR spectroscopy

Citation: Santanni, F.; Chelazzi, L.; Sorace, L.; Timco, G.A.; Sessoli, R. Structural and Magnetic Properties of the {Cr(pybd)$_3$[Cu(cyclen)]$_2$}(BF$_4$)$_4$ Heteronuclear Complex. *Crystals* **2023**, *13*, 901. https://doi.org/10.3390/cryst13060901

Academic Editor: Younes Hanifehpour

Received: 2 May 2023
Revised: 22 May 2023
Accepted: 27 May 2023
Published: 1 June 2023

Copyright: © 2023 by the authors. Licensee MDPI, Basel, Switzerland. This article is an open access article distributed under the terms and conditions of the Creative Commons Attribution (CC BY) license (https://creativecommons.org/licenses/by/4.0/).

1. Introduction

The qubit, the quantum analog of classical bits, bases its working principles on quantomechanical laws such as the superposition of states. In principle, it is possible to exploit two spin levels of an unpaired electron as the computational basis and encode the information in linear combinations of these states [1]. Magnetic molecules characterized by good spin coherence properties are appealing candidates for improving the quantum logical units of quantum computers [2,3]. They can work as qubits when two electronic spin levels are employed to encode the information (e.g., the $m_S = \pm 1/2$ projections along the quantization axis of an S = 1/2 system) [4,5] or as qudits when more electronic and nuclear spin levels are involved [6,7].

When connected through specific channels (e.g., superconductive planar waveguides [8]), multi-qubit architectures capable of implementing more complex logic operators known as quantum logic gates can be achieved [9]. Likewise, it is possible to "connect" different molecular spins to mime such architectures present in modern quantum processors [10].

Beyond the tailoring of organic ligands and molecules to obtain multi-spin structures [11,12], many other conditions must be fulfilled. For instance, most logical operations and quantum algorithms require that qubits can act as entangled objects. In other words, an interaction between the two spins must be active, and the total state cannot be expressed as a simple product of the individual spin states. Besides this, entangled qubits must be individually addressable so that the selective manipulation of qubits results in the detectable variation of the counterparts quantum states. Magnetic interactions between the spin centers, dipolar or isotropic exchange in nature [11–14], can be exploited to achieve the required entanglement condition [15]. However, the magnetic coupling interaction should be small enough, typically of the order of 10^{-1}–10^{-2} cm^{-1}, to avoid the formation of a giant spin state. Two main strategies can be used to maintain the individual addressability of spins: (i) the introduction of specific ligands and organic scaffolds that guarantee a different relative orientation of the magnetic anisotropy tensors of spin centers [11,16] or (ii) the exploitation of different metal or spin centers within a single architecture, these being characterized by different magnetic anisotropy tensors [17–19].

Heterometallic coordination compounds are of high interest in various fields, spanning from catalysis [20–23] to molecular magnetism [24–28]. Obtaining such structures is not a trivial task, and a tailored design of employed ligands, eventually heterotopic ones [29–31], is often necessary. The use of heterometallic structures has been proposed as a potential strategy to implement quantum logic gates. As reported in [12], the entanglement condition among the different qubits within these architectures can be selectively enabled upon applying specific pulse sequences. In this sense, such a strategy is preferable to the first one since it allows us to mime what commonly happens in quantum circuits, where qubits act as single objects or entangled units upon necessity. However, two critical aspects can be identified for this strategy: (i) the obtainment of heterometallic multi-qubit systems is not trivial, nor are the design and synthetic efforts required [32]; (ii) it is difficult to predict the final quantum and magnetic properties of adducts, where parties can behave differently compared to their isolated forms.

Among the most investigated systems, molecular metallocages represent a significant part of reported systems [31,33,34]. Starting from the works of Brechin and coworkers on Cr-based cages built by the (1-pyridyl)-butane-1,3-dione (Hpybd) ligand [30,35–37], we focused on obtaining a heterometallic system by using the same ligand. Such a ligand allows for getting octahedral complexes of M^{3+} ions thanks to the coordinating nature of the diketonate core. Then, by further exploiting the peripheral pyridyl functions, it is also possible to coordinate up to three additional ions per M^{3+} unit. In our case, we exploited the ditopic nature of this ligand to obtain a heterometallic complex containing a central Cr^{3+} ion and Cu^{2+}-based peripheral units. Differently from previous reports [30,35], we exploited the pyridyl units to coordinate a Cu^{2+} complex, i.e., the [Cu(cyclen)]$^{2+}$ (cyclen = 1,4,7,10-tetraazacyclododecane). Using the cyclen moiety as a "capping agent" allowed us to get a finite-size complex that avoids forming polymers or cage-like structures.

In this work, we report the structural characterization of the {Cr(pybd)$_3$[Cu(Cyclen)]$_2$}(BF$_4$)$_4$ compound followed by powder X-ray diffraction studies and data analysis. Furthermore, we report the investigation of the static magnetic properties of the adduct performed by standard magnetometry and electron paramagnetic resonance (EPR) spectroscopy.

2. Materials and Methods

2.1. Synthesis

2.1.1. General Remarks

All the reagents and solvents were of reagent grade and employed without further purification. Compounds Hpybd and [Cr(pybd)$_3$] were synthesized according to the literature [30]. More details about the synthesis of all the other compounds are reported in Section S1 of Supplementary Information (SI). A general reaction pathway is reported in Scheme 1 for clarity.

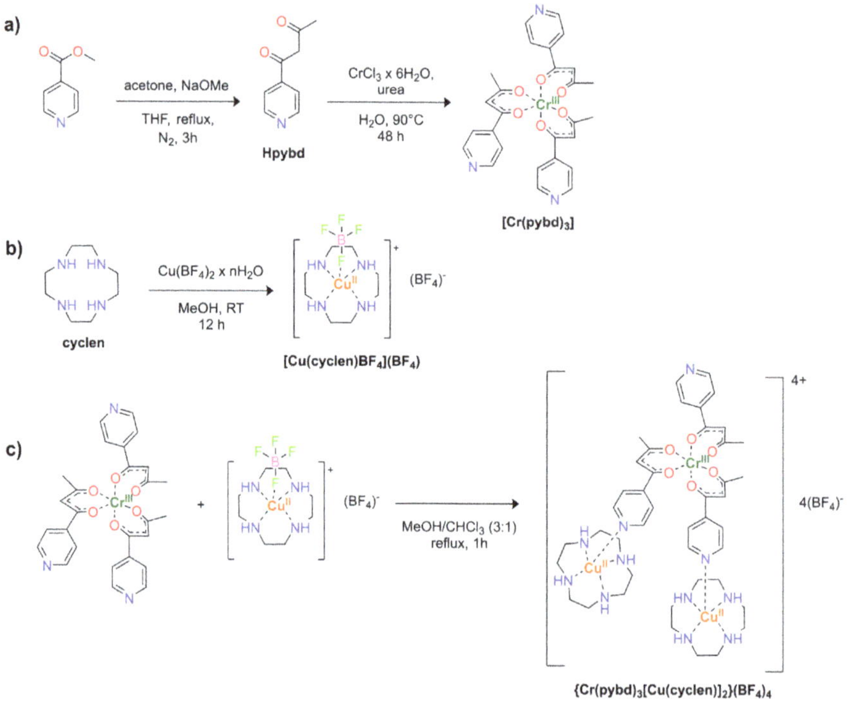

Scheme 1. Synthetic schemes showing the general strategies employed for the obtaining of reported compounds. (**a**) Synthesis of Hpybd and [Cr(pybd)₃]. (**b**) Synthesis of [Cu(cyclen)BF₄](BF₄). (**c**) Synthesis of the macromolecular adduct {Cr(pybd)₃[Cu(cyclen)]₂}(BF₄)₄.

2.1.2. Synthesis of {Cr(pybd)$_3$[Cu(cyclen)]$_2$}(BF$_4$)$_4$

In a 50 mL two-neck flask equipped with a condenser, [Cu(cyclen)BF$_4$]BF$_4$ (235 mg, 0.6 mmol) was dissolved in 15 mL of MeOH and heated to reflux under magnetic stirring. A solution of [Cr(pybd)$_3$] (100 mg, 0.2 mmol) in CHCl$_3$ (5 mL) was added dropwise to the reaction mixture, and the solution refluxed for 1 h. Then, the solution was cooled to room temperature, filtered, and evaporated slowly. After three days, a crystalline product (218 mg, 85%) was collected by filtration and washed with CHCl$_3$ (5 mL × 2) and Et$_2$O (5 mL × 2). The grounded powder was employed for structural determination (see below). Elemental analysis (calc. for CrCu$_2$C$_{43}$H$_{64}$N$_{11}$O$_6$B$_4$F$_{16}$): C, 38.45 (38.05); H, 4.37 (4.75); N; 11.29 (11.35). UV/Vis (λ_{max}, nm): 279, 368, 587 (ε = 43 M^{-1} cm^{-1}); see also Figure S1. FT-IR (ν, cm^{-1}): 3325(w), 3306(s), 3129(vw), 2941(w), 2895(w), 1585(m), 1545(w), 1515(s), 1495(vw), 1429(m), 1394(s), 1322(vw), 1291(m), 1250(vw), 1217(w), 1081 (vs, b), 1057(vs, b) 974(m, b), 955(vw, b), 867(w), 849(w), 810(vw), 775(m), 701(s), 619(s) (Figure S2 of SI).

2.2. *UV–Vis Spectroscopy*

UV–Vis spectra were recorded on a Jasco V-670 double-beam spectrophotometer by using quartz cuvettes of 1 cm length. All experiments were performed on 0.25 M MeOH solutions of employed compounds. All experimental spectra are reported in Figures S1 and S3 of SI.

2.3. *X-ray Structural Characterization*

Single-crystal X-ray diffraction was employed to obtain the unit cell parameters. A single crystal was analyzed at 100 K using Cu-Kα radiation (λ = 1.54184 Å) on a Bruker Apex-II diffractometer equipped with pixel array detector PhotonII, controlled using APEX2 software [38].

Powder X-ray Diffraction (PXRD) was used for structure determination using unit cell parameters extracted from single crystal data. High-quality PXRD data were recorded in a 0.3 mm glass capillary at room temperature by using a Bruker New D8 Da Vinci diffractometer (Cu − Kα1 radiation = 1.54056 Å, 40 kV × 40 mA) equipped with a Bruker LYNXEYE-XE detector, scanning range 2θ = 3–70°, 0.02° increments of 2θ, and a counting time of 576 total time/step. Space group determination with EXPO2014 [39] resulted in space group P-1 with Z = 2. We then solved the structure by stimulated annealing using the EXPO2014 suite. Each annealing trial works on three runs, using a cooling rate (T_n/T_{n-1}) of 0.95. All the torsion angles were allowed to rotate freely during the refinement process while bond distances and angles were kept fixed. The best solution was chosen for Rietveld refinement, which was performed with the software TOPASv6 [40]. Background and peak shape were fitted using a shifted Chebyshev function with eight coefficients and a Pseudo-Voigt function, respectively. All atoms were isotropically refined with a typical thermal parameter depending on the element. All the hydrogen atoms were fixed in calculated positions. Crystal data and refinement parameters are reported in Table 1. The simulated pattern was obtained by using the software Mercury CSD 2022.2.0 (copyright CCDC, https://www.ccdc.cam.ac.uk/solutions/software/mercury, accessed: 26 May 2023) [41] by employing an FWHM value of 0.1 and an increment step size of 0.025°. Complete crystallographic data for the solved structures have been deposited in the Cambridge Crystallographic Data Centre with CCDC number 2260168.

Table 1. Crystallographic data and refinement parameters.

Molecular Formula	$C_{42}CrCu_2N_{12}O_6$,4(BF_4)
Mr	1294.8
T(K)	RT
λ (Å)	1.54056
crystal system, space group	triclinic, P-1
unit cell dimensions (Å, °)	a = 12.5014(5), b = 15.2007(8), c = 18.4058(8) α = 105.05(4), β = 90.228(5), γ = 95.988(3)
volume (Å^3)	3357.5(3)
Z, Dx (g/cm^{-3})	2, 1.325
μ (mm^{-1})	0.906
Rwp (%)	4.66
GOOFs	1.52

2.4. DC Magnetometry

Direct current (DC) magnetic measurements were performed in the temperature range 2–300 K (B = 0.1 T) or in the field range 0–5 T (T = 2 K, 4 K, 8 K) using a Quantum Design Magnetic Properties Measuring System (QD-MPMS) equipped with a Superconductive Quantum Interference Device (SQUID). Magnetization data were corrected for the diamagnetic contributions of sample holders (previously measured using the same conditions) and of the sample itself, as deduced by using Pascal's constant tables [42].

2.5. Electron Paramagnetic Resonance (EPR) Spectroscopy

CW X-band EPR spectra of all samples were recorded on a Bruker Elexsys E500 spectrometer equipped with an SHQ cavity (ν = 9.43 GHz). Low-temperature measurements were obtained using an Oxford Instruments ESR900 continuous-flow helium cryostat. All measured samples were 0.5 mM solutions of employed compounds in EtOH/MeOH 4:1. The solutions were introduced within a standard X-band quartz tube and rapidly frozen in a liquid nitrogen bath to obtain dispersions in glassy matrixes.

3. Results and Discussion

3.1. Synthesis

The molecular complex [Cr(pybd)$_3$] was synthesized following the strategy by Brechin et al. [30,35–37]. The reaction of CrCl$_3$·6H$_2$O with the Hpybd ligand was performed in water by using urea as the base for deprotonating the employed diketonate. Such a strategy is commonly used to avoid the formation of chromium oxides and hydroxides during the reaction since ammonia slowly develops by urea decomposition at 90 °C [43]. After purification, the [Cr(pybd)$_3$] complex was reacted with three equivalents of the Cu^{2+} precursor: [Cu(cyclen)BF$_4$]BF$_4$, in a MeOH/CHCl$_3$ 6:1 solution. This procedure was used with the intent to introduce three Cu units per Cr center, as done in ref. [30], and to obtain the tetramer {CrCu$_3$}. The reaction can be followed by observing the color changing from royal blue to greenish/blue. On cooling, dark crystals formed from the solution, and larger amounts of the product were obtained by slow evaporation of the reaction mixture. Notwithstanding our predictions and efforts to obtain a tetramer, it was only possible to get the trimeric specie {Cr(pybd)$_3$[Cu(cyclen)]$_2$}(BF$_4$)$_4$ (see below). Interestingly, the obtained compound does not respect the expected Cr/Cu 1:3 stoichiometry even though a large excess of Cu^{2+} precursor is employed. This behavior could be imputed to the steric hindrance introduced by adding [Cu(cyclen)]$^{2+}$ to the pyridyl units. Furthermore, it was only possible to obtain {Cr(pybd)$_3$[Cu(cyclen)]$_2$}(BF$_4$)$_4$ by slow evaporation of the reaction mixture. Any other attempt to recrystallize it from solutions of different organic solvents and their mixtures (e.g., MeOH, CHCl$_3$, DMSO, and DMF) led to the separation of individual Cr^{3+} and Cu^{2+} precursors.

3.2. Structural Characterization

As mentioned above, {Cr(pybd)$_3$[Cu(cyclen)]$_2$}(BF$_4$)$_4$ (Figure 1) crystallizes from the reaction mixture upon cooling and subsequent evaporation of MeOH. Nevertheless, the plate-shaped crystals tend to pile along the *c* crystallographic axis. For this reason, cutting a single crystal suitable for X-ray structural determination was not possible. Hence, we employed the single-crystal diffraction tool to firstly estimate the cell parameters that were successively used for the structural refinement model employed on PXRD data (see Section 2). We report the comparison between the experimental and simulated PXRD patterns in Figure 2 as proof of the quality of the employed model. The plot evidences a remarkable agreement between experiment and simulation, both in terms of peak positions and relative intensities.

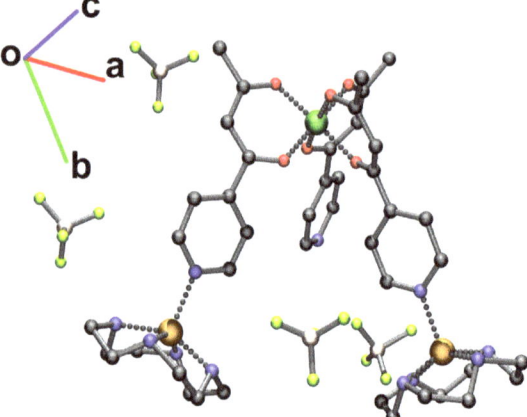

Figure 1. View of the molecular structure of {Cr(pybd)$_3$[Cu(cyclen)]$_2$}(BF$_4$)$_4$. Hydrogen atoms are omitted for clarity. Color code: Cr = green; Cu = copper; F = yellow/green; O = red; N = blue; C = grey; B = pink.

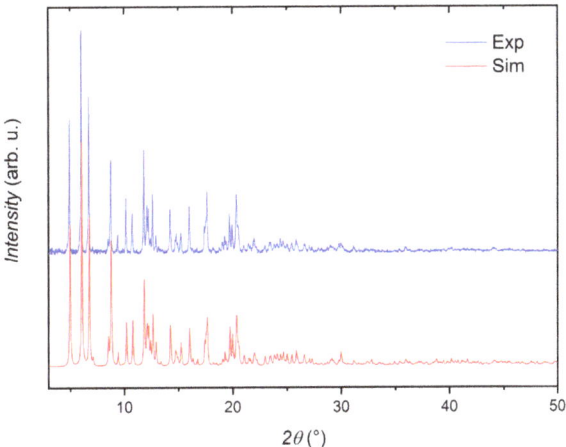

Figure 2. Comparison between experimental and simulated PXRD patterns obtained for microcrystalline {[Cr(pybd)$_3$][Cu(cyclen)]$_2$}(BF$_4$)$_4$.

{[Cr(pybd)$_3$][Cu(cyclen)]$_2$}(BF$_4$)$_4$ crystallizes in the P−1 space group (no. 2) with two {[Cr(pybd)$_3$][Cu(cyclen)]$_2$}$^{4+}$ cationic species and eight BF$_4^-$ counterions per unit-cell. The asymmetric unit is represented by the {[Cr(pybd)$_3$][Cu(cyclen)]$_2$}(BF$_4$)$_4$ unit reported in Figure 1 (unit cell content shown in Figure S3 of SI). The crystal structure also presents solvent-accessible pores, occupying 5% of the unit cell volume with a volume of about 165 Å^3 (Figure S4 of SI). The porous structure propagates along the *a*-axis direction.

In the asymmetric unit, the Cr^{3+} ion preserves the *fac* isomerism typical of [Cr(pydb)$_3$] [30] and presents a distorted octahedral coordination geometry having mean Cr−O distances of 1.96 Å on the equatorial plane and Cr−O5 and Cr−O6 distances of 1.93 Å and 1.97 Å along the axial direction (see Figure S5a of SI). Furthermore, all the O2−Cr−O4, O1−Cr−O3, and O5−Cr−O6 angles are smaller than 180°, being of about 176.0°, 177.0°, and 179.0°, respectively (see Figure S5b of SI). Such a distorted octahedral geometry differs from the regular one reported for the [Cr(pydb)$_3$] precursor, although it has been observed for [Cr(pydb)$_3$]-based supramolecular cages [30,35,36]. In analogy to the Cr^{3+} ion, the two Cu^{2+} moieties present slightly distorted square-pyramidal geometry due to minor differences in the Cu−N$_{cyclen}$ bond lengths (see Figure S6 of SI). In this case, the mean Cu−N$_{cyclen}$ distance is 2.03 Å for both cyclen based units, while the Cu−N$_{py}$ distances are 2.15 Å and 2.17 Å for Cu1−N5 and Cu2−N10, respectively. These distances align with other square-pyramidal N-based copper complexes [44]. The mean Cr−Cu intramolecular distance is 8.9 Å. The smallest Cu···Cu intermolecular distance of 7.3 Å is measured between molecules in adjacent cells moving along the *b*-axis. The smallest intermolecular Cr···Cr distance is 7.36 Å, and it is measured between molecules in adjacent cells along the *c* axis. The structure presents four BF$_4^-$ per molecular unit, three of them presenting short-contact interactions with the acidic N−H moieties of the two [Cu(cyclen)]$^{2+}$ units per molecule. At the same time, the fourth is stabilized within the framework by similar interactions set with the cyclen unit of another molecule.

3.3. UV–Vis Spectroscopy Characterization

To check for the stability of the {[Cr(pybd)$_3$][Cu(cyclen)]$_2$}$^{4+}$ adduct when in solution and confirm the coordination geometry of metal sites, we performed a UV–Vis spectroscopic investigation on 2.5 × 10^{-3} M solutions of [Cu(cyclen)]$^{2+}$ and {[Cr(pybd)$_3$][Cu(cyclen)]$_2$}$^{4+}$ in MeOH and compared their experimental spectra. To do that, we focused on the spectral region between 400 and 850 nm, where electronic transitions attributed to the copper(II) center are usually observed [44–47]. The graph in Figure 3 shows that the two complexes present an absorption band centered at about 600 nm. In particular, we observed absorption maxima at 587 nm (ε = 43 M^{-1} cm^{-1}) and 601 nm (ε = 35 M^{-1} cm^{-1}) for {[Cr(pybd)$_3$][Cu(cyclen)]$_2$}$^{4+}$ and [Cu(cyclen)]$^{2+}$, respectively. This spectral feature is imputed to a d-d transition of the copper center, and the observed energy is characteristic of pentacoordinate complexes having a square-pyramidal coordination geometry [44]. We further compared these UV–Vis spectra with an additional one obtained on a 2.5 × 10^{-3} M solution of [Cu(cyclen)]$^{2+}$ in MeOH added in 100 equivalents of pyridine. This third spectrum shows a red shift of the absorption maximum to λ = 617 nm (ε = 35 M^{-1} cm^{-1}), which is expected for stronger interactions with an axial ligand on the fifth position [48]. By comparing this spectrum with that obtained on {[Cr(pybd)$_3$][Cu(cyclen)]$_2$}$^{4+}$, we can state that the coordination of the Cu^{2+} ion by the pyridyl unit of pybd ligand leads to a smaller crystal field interaction with respect to pure pyridine. This can be correlated to a weaker interaction with the unpaired doublet of the N donor atom, whose delocalization on the conjugated β-diketonate scaffold could be further affected by the electron-withdrawing effect of the coordinated Cr^{3+} ion.

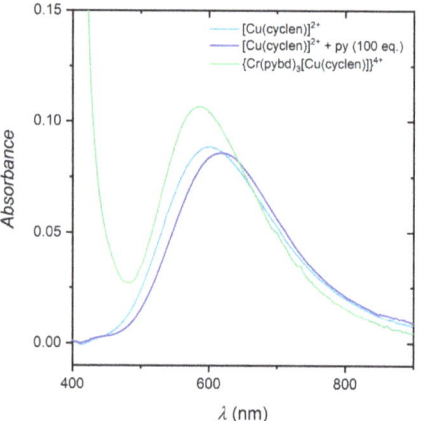

Figure 3. UV–Vis absorption spectra collected on 2.5 × 10^{-3} M MeOH solutions of [Cu(cyclen)]$^{2+}$, [Cu(cyclen)]$^{2+}$ + 100 eq of pyridine (py), and {Cr(pybd)$_3$[Cu(cyclen)]$_2$}$^{4+}$.

3.4. Magnetism and EPR Spectroscopy

X-band ($\nu \approx$ 9.4 GHz) continuous wave EPR measurements were performed to characterize and compare the magnetic anisotropy properties of the presented systems. To do that, we measured frozen solution of [Cr(pybd)$_3$], [Cu(cyclen)]$^{2+}$, and {Cr(pybd)$_3$[Cu(cyclen)]$_2$}$^{4+}$ in EtOH/MeOH 4:1 (see Section 2). The experimental spectra are reported as blue lines in Figure 4a–c.

The [Cr(pybd)$_3$] complex shows the characteristic EPR spectrum of an octahedrally coordinated Cr^{3+} ion (Figure 4a). The observed spectral features are attributed to spin transitions within the quartet basis state (S = 3/2) further split in two Kramers' doublets ($m_S = \pm 1/2$; $m_S = \pm 3/2$) by the zero-field splitting (ZFS) interaction, which is characterized by axial (D) and rhombic (E) terms following the spin Hamiltonian (SH) in Equation (1).

$$\hat{H}_S^{Cr} = -\mu_B \hat{S} \cdot g \cdot \vec{B} + D\hat{S}_z^2 + E\left(\hat{S}_x^2 - \hat{S}_y^2\right) \quad (1)$$

The spectrum can be well simulated [49] with the SH parameters extracted from ref. [30] and reported in Table 1. The chromium complex is characterized by an easy-plane anisotropy (D = + 0.55 cm^{-1}) leading to $m_S = \pm 1/2$ ground doublet. The E/D ratio of 0.045 confirms that the system has a weak rhombic distortion. Such values are in line with what has been reported in the literature for other Cr-acetylacetonate complexes, for which D values of about 0.6 cm^{-1} have been observed [50,51].

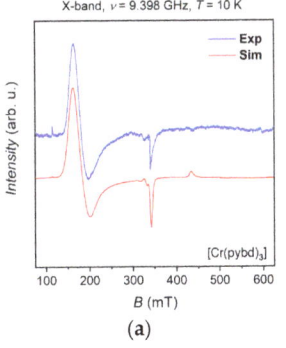 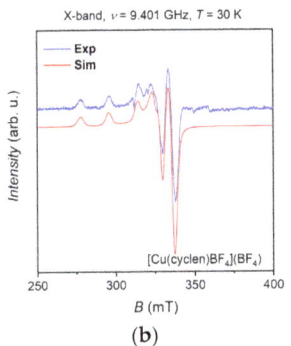 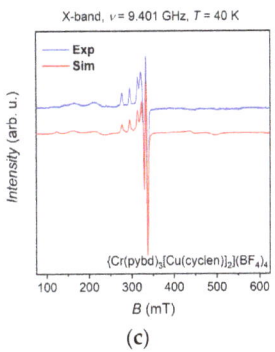

(a) (b) (c)

Figure 4. (**a**) Experimental and simulated X-band EPR spectra of [Cr(pybd)$_3$] 0.5 mM in EtOH/MeOH 4:1; (**b**) Experimental and simulated X-band EPR spectra of [Cu(cyclen)]$^{2+}$ 0.5 mM in EtOH/MeOH 4:1; (**c**) Experimental and simulated X-band EPR spectra of {Cr(pybd)$_3$[Cu(cyclen)]$_2$}$^{4+}$ 0.5 mM in EtOH/MeOH 4:1.

The [Cu(cyclen)]$^{2+}$ compound showed the EPR spectrum in Figure 4b, which is closely in line with that expected for an $S = 1/2$ system characterized by an anisotropic axial g tensor with an axial hyperfine coupling interaction (A) between the spin and the copper nucleus ($I = 3/2$). A quantitative estimation of the magnetic anisotropy followed from the simulation of the experimental spectrum with the SH reported in Equation (2).

$$\hat{H}_S^{Cu} = -\mu_B \cdot \hat{S} \cdot g \cdot \vec{B} + \hat{I} \cdot A \cdot \hat{S} \qquad (2)$$

The values obtained for the components of g and A tensors (Table 2), for which $g_{||} > g_\perp$ and $A_{||} > A_\perp$ (with $z = ||$ and $x = y = \perp$), are representative of a pentacoordinate Cu^{2+} complex [47,48] and confirm the experimental evidence given by the UV–Vis analysis (see above).

Table 2. List of best-simulation parameters extracted for X-band spectra of 0.5 mM frozen solutions in EtOH/MeOH 4:1. In the case of the Cr-Cu adduct, we did not consider the relative orientation of magnetic reference frameworks of each site.

Parameters	[Cr(pybd)$_3$]	[Cu(cyclen)BF$_4$](BF$_4$)	{Cr(pybd)$_3$[Cu(cyclen)]}(BF$_4$)$_4$	
g_x	1.97(1)	2.047(2)	1.97(1) [a]	2.050(2) [b]
g_y	1.97(1)	2.047(2)	1.97(1) [a]	2.050(2) [b]
g_z	1.97(1)	2.204(2)	1.97(1) [a]	2.206(2) [b]
D (cm^{-1})	0.55(5)	-	0.21(1) [a]	-
E (cm^{-1})	0.025(1)	-	0.02(1) [a]	-
A_x (10^{-4} cm^{-1})	-	17(2)	-	18(2) [b]
A_y (10^{-4} cm^{-1})	-	17(2)	-	18(2) [b]
A_z (10^{-4} cm^{-1})	-	182(2)	-	184(2) [b]

a = set of parameters extracted for the Cr^{3+} ion; b = set of parameters extracted for the two equivalent Cu^{2+} sites.

The X-band characterization of {Cr(pybd)$_3$[Cu(cyclen)]$_2$}$^{4+}$ was performed at 40 K to avoid saturation of the Cu^{2+} signal and simultaneously detect that of the Cr^{3+} ion. The obtained spectrum (Figure 4c) slightly differs from the trivial sum of those of the two precursors. On the one hand, the low-field region between 100 mT and 25 mT, i.e., the most indicative one for the Cr^{3+} ion signal, largely differs from that of isolated [Cr(pybd)$_3$]. On the other hand, the Cu^{2+} signal falling within the 250–400 mT region seems not to be affected by the presence of the Cr^{3+} ion. Based on this observation, we can say that there is no evidence of a clear isotropic exchange coupling interaction among the Cr and Cu sites. It follows that the unique modification of the chromium signal should be correlated to the variation of the SH parameters of the Cr^{3+} ion upon copper coordination of two pyridine rings of the pybd ligand, while those of Cu^{2+} remain unaltered. To validate such an assumption, we simulated the spectra by fixing the SH parameters of Cu^{2+} moieties and varying the D and E components of chromium using a SH given by the sum of those reported in Equations (1) and (2). The best simulation reported in Figure 4c was obtained by reducing both the D and E values to 0.21 cm^{-1} and 0.02 cm^{-1}, respectively.

Consequently, the E/D ratio of the adduct increased to 0.095, highlighting a larger rhombicity of the system. This feature can be correlated to the distortion of the Cr^{3+} coordination sphere observed in the case of {Cr(pybd)$_3$[Cu(cyclen)]$_2$}$^{4+}$ (see Structural Characterization section), which led to an increased rhombicity. It must be noted that, in our model, dipolar interactions among spins were not introduced directly, but they were included as contributions to the isotropic and anisotropic line-broadening. These interactions could be better highlighted and investigated by adopting advanced pulsed EPR experiments such as DEER (Double Electron-Electron Resonance) [52].

To further exclude that the lack of magnetic exchange interaction is not a consequence of structural changes occurring in the frozen solution, we measured crystalline powders of {Cr(pybd)$_3$[Cu(cyclen)]$_2$}(BF$_4$)$_4$. However, the EPR spectrum collected on the solid sample is significantly broadened because of intermolecular interactions (Figure S7 of SI), and it cannot pursue our scope. For this reason, we employed DC magnetometry (see Section 2) to get more insights into the magnetic interactions characteristic of the compound. The magnetic susceptibility (χ) data are reported in Figure 5a as χT product vs. T. Above 25 K, the χT value reaches a plateau around 2.62 cm^3 mol^{-1} K, which is the sum of the Curie's constants expected for the three paramagnetic centers characterized by the ZFS and **g** tensors anisotropies determined by our EPR experiments. Below this temperature, the χT value drops abruptly to 2.4 cm^3 mol^{-1} K. This behavior can be correlated to two distinct factors, i.e., (i) a contribution linked to the ZFS of Cr^{3+} ion and (ii) a contribution due to intra- or intermolecular coupling interactions within the crystal lattice. In order to estimate the magnitude of such interactions, we performed a simultaneous fit of χT vs. T and M vs. B (Figure 5a,b) curves with the software PHI [53]. We used the set of SH parameters extracted from X-band measurements and kept them fixed, while the mean-field intermolecular interactions parameter (zJ) was used as the fitting parameter. The employed model provides a best-fit zJ value of $1.92(1) \cdot 10^{-2}$ cm^{-1}. We repeated the same fitting procedure by replacing the mean-field interaction with an equivalent isotropic exchange coupling interaction ($H_{ex} = -J_{Cr-Cu}(\hat{S}_{Cu1} \cdot \hat{S}_{Cr} + \hat{S}_{Cu2} \cdot \hat{S}_{Cr})$) between the two Cu^{2+} ions and the central Cr^{3+} and fitting the coupling constant J_{Cr-Cu}. The best $J_{Cr-Cu} = -5.71(1) \cdot 10^{-2}$ cm^{-1} is of the same order of magnitude as the mean-field constant zJ, and the fitted curve (red line in Figure 5) poorly reproduces the experimental χT points between 2 and 5 K.

This latter analysis, together with our EPR results reported for the diluted phase, suggests that, even if present, exchange coupling interactions are not detectable in the limit of the experimental conditions employed.

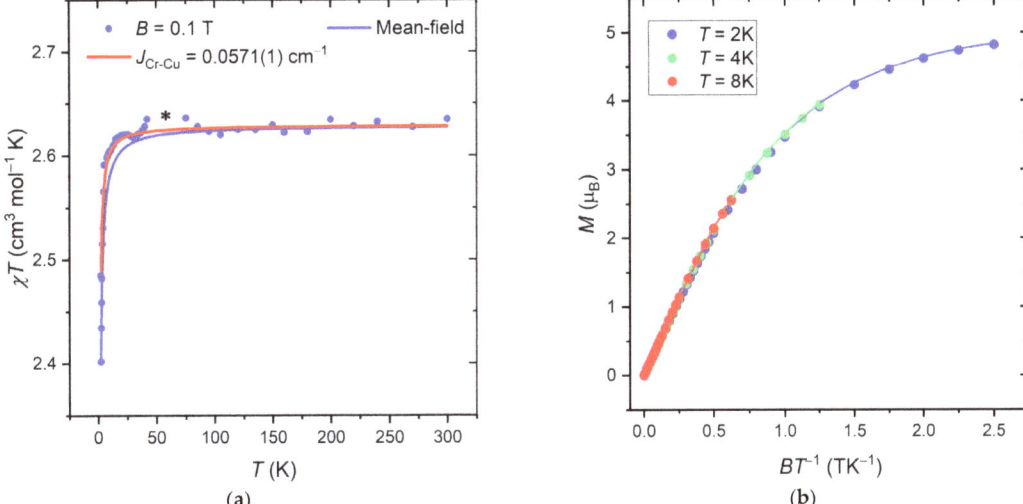

Figure 5. (a) χT vs. T plot obtained for {[Cr(pybd)$_3$][Cu(cyclen)]$_2$}(BF$_4$)$_4$ at $B = 0.1$ T. An asterisk is placed in correspondence of a signal anomaly due to oxygen impurity within the measurement chamber; (b) M vs. BT^{-1} plot obtained for {[Cr(pybd)$_3$][Cu(cyclen)]$_2$}(BF$_4$)$_4$ at $T = 2, 4,$ and 8 K. Solid lines within the plot correspond to the fitting outcomes.

4. Conclusions

In this work, we reported the synthesis of the new molecular cluster {[Cr(pybd)$_3$][Cu(cyclen)]$_2$}(BF$_4$)$_4$ by exploiting the heterotopic properties of the pybd$^-$ ligand and using cyclen rings as capping agents to avoid the formation of molecular cages. The crystallographic structure has been resolved thanks to a combined approach involving single-crystal X-ray measurements and powder X-ray diffraction suites. Congruently to our elemental analysis results, the crystallographic characterization unveiled the structural peculiarity of the cluster, which is characterized by the presence of just two coordinated [Cu(cyclen)]2+ moieties over the three potential pyridyl units of the [Cr(pybd)$_3$] core. Our X-band CW-EPR measurements on diluted frozen solutions highlighted the variation of the magnetic anisotropy properties of the Cr^{3+} center in {[Cr(pybd)$_3$][Cu(cyclen)]$_2$}(BF$_4$)$_4$ with respect to the monomeric [Cr(pybd)$_3$] unit. The increased rhombic distortion in {[Cr(pybd)$_3$][Cu(cyclen)]$_2$}(BF$_4$)$_4$, highlighted by an E/D ratio of $9.5 \cdot 10^{-2}$, can be justified on the basis of crystallographic results, which pinpoints the more significant distortion of the Cr^{3+} coordination geometry in the trimer compared to the monomer. Furthermore, both EPR and magnetometry measurements highlighted a small coupling interaction between the central spin and the peripheral units, primarily relying on a dipolar coupling interaction between paramagnetic centers. This result is in contrast with what has been observed for Cu-Cr macromolecular cages obtained with the same [Cr(pybd)$_3$] building block, for which a ferromagnetic coupling interaction of about 10^{-1} cm^{-1} has been reported [30,35]. However, while the Cu^{2+} ions included within cages are coordinated on equatorial positions by pyridyl units of the pybd$^-$ ligand, the same units point toward the axial position of the pentacoordinate [Cu(cyclen)]$^{2+}$ units present in our cluster. In this respect, we could have expected a lower super-exchange interaction between the Cr^{3+} and Cu^{2+} ions in our cluster than in that of ref. [30] since here, the pyridyl unit points towards the doubly occupied d_{z^2} of the copper ion, rather than the singly occupied $d_{x^2-y^2}$. Unfortunately, the change in the orbital involved in the interaction reduces the interaction to a value smaller than the dipolar ones that is not detectable within the employed experimental conditions.

Supplementary Materials: The following supporting information can be downloaded at: https://www.mdpi.com/article/10.3390/cryst13060901/s1, more details about syntheses and employed chemicals; Figure S1: UV–Vis absorption spectra of equimolar solutions (0.25 mM) of all the investigated complexes; Figure S2: FT-IR spectra collected on [Cr(pybd)$_3$], [Cu(cyclen)BF$_4$](BF$_4$), and {Cr(dypb)$_3$[Cu(cyclen)]}(BF$_4$)$_2$ powder samples; Figure S3: View of the crystal unit cell of {Cr(dypb)$_3$[Cu(cyclen)]}(BF$_4$)$_2$ along with the a, b, and c crystallographic axes; Figure S4: View along the c axis of a 2 × 2 × 1.5 cell highlighting the porous nature of crystalline {Cr(dypb)$_3$[Cu(cyclen)]}(BF4)$_2$; Figure S5: (a) Generic view of the first coordination sphere of the Cr^{3+} ion reporting the six values of Cr−O distances. (b) Generic view of the first coordination sphere of the Cr^{3+} io reporting the three values of O−Cr−O angles; Figure S6: Generic view of the two Cu(cyclen) moieties highlighting the square-pyramidal coordination environment of the Cu^{2+} ion; Figure S7: X-band CW EPR spectrum of a {Cr(dypb)$_3$[Cu(cyclen)]}(BF4)$_2$ powder sample collected at 30 K.

Author Contributions: Conceptualization, F.S. and R.S.; formal analysis, F.S., L.C., G.A.T., L.S. and R.S.; investigation, F.S. and L.C.; data curation, F.S., L.C., G.A.T., L.S. and R.S.; writing—original draft preparation, F.S. and R.S.; writing—review and editing, F.S., L.C., G.A.T., L.S. and R.S.; supervision, L.S., G.A.T. and R.S.; project administration, R.S.; funding acquisition, R.S. All authors have read and agreed to the published version of the manuscript.

Funding: The support of EC through the EU Commission through the FETOPEN project FAT-MOLS (GA 862893) and Italian MIUR through Progetto Dipartimenti di Eccellenza 2023–2027 (CUP B97G22000740001 - DICUS 2.0) to the Department of Chemistry "Ugo Schiff" of the University of Florence and PNRR MUR project PE0000023-NQSTI is acknowledged.

Data Availability Statement: Data are available from authors upon request.

Conflicts of Interest: The authors declare no conflict of interest.

References

1. Heinrich, A.J.; Oliver, W.D.; Vandersypen, L.M.K.; Ardavan, A.; Sessoli, R.; Loss, D.; Jayich, A.B.; Fernandez-Rossier, J.; Laucht, A.; Morello, A. Quantum-Coherent Nanoscience. *Nat. Nanotechnol.* **2021**, *16*, 1318–1329. [CrossRef]
2. Atzori, M.; Sessoli, R. The Second Quantum Revolution: Role and Challenges of Molecular Chemistry. *J. Am. Chem. Soc.* **2019**, *141*, 11339–11352. [CrossRef] [PubMed]
3. Gaita-Ariño, A.; Luis, F.; Hill, S.; Coronado, E. Molecular Spins for Quantum Computation. *Nat. Chem.* **2019**, *11*, 301–309. [CrossRef]
4. Bonizzoni, C.; Ghirri, A.; Santanni, F.; Atzori, M.; Sorace, L.; Sessoli, R.; Affronte, M. Storage and Retrieval of Microwave Pulses with Molecular Spin Ensembles. *npj Quantum Inf.* **2020**, *6*, 68. [CrossRef]
5. Bonizzoni, C.; Tincani, M.; Santanni, F.; Affronte, M. Machine-Learning-Assisted Manipulation and Readout of Molecular Spin Qubits. *Phys. Rev. Appl.* **2022**, *18*, 064074. [CrossRef]
6. Chicco, S.; Chiesa, A.; Allodi, G.; Garlatti, E.; Atzori, M.; Sorace, L.; De Renzi, R.; Sessoli, R.; Carretta, S. Controlled Coherent Dynamics of [VO(TPP)], a Prototype Molecular Nuclear Qudit with an Electronic Ancilla. *Chem. Sci.* **2021**, *12*, 12046–12055. [CrossRef]
7. Chizzini, M.; Crippa, L.; Zaccardi, L.; Macaluso, E.; Carretta, S.; Chiesa, A.; Santini, P. Quantum Error Correction with Molecular Spin Qudits. *Phys. Chem. Chem. Phys.* **2022**, *24*, 20030–20039. [CrossRef]
8. Hazra, S.; Bhattacharjee, A.; Chand, M.; Salunkhe, K.V.; Gopalakrishnan, S.; Patankar, M.P.; Vijay, R. Ring-Resonator-Based Coupling Architecture for Enhanced Connectivity in a Superconducting Multiqubit Network. *Phys. Rev. Appl.* **2021**, *16*, 024018. [CrossRef]
9. Nielsen, M.A.; Chuang, I.L. *Quantum Computation and Quantum Information*; Cambridge University Press: Cambridge, UK, 2010; ISBN 9780511976667.
10. Lockyer, S.J.; Chiesa, A.; Timco, G.A.; McInnes, E.J.L.; Bennett, T.S.; Vitorica-Yrezebal, I.J.; Carretta, S.; Winpenny, R.E.P. Targeting Molecular Quantum Memory with Embedded Error Correction. *Chem. Sci.* **2021**, *12*, 9104–9113. [CrossRef]
11. Ranieri, D.; Santanni, F.; Privitera, A.; Albino, A.; Salvadori, E.; Chiesa, M.; Totti, F.; Sorace, L.; Sessoli, R. An Exchange Coupled Meso—Meso Linked Vanadyl Porphyrin Dimer for Quantum Information Processing. *Chem. Sci.* **2023**, *14*, 61–69. [CrossRef]
12. Ferrando-Soria, J.; Moreno Pineda, E.; Chiesa, A.; Fernandez, A.; Magee, S.A.; Carretta, S.; Santini, P.; Vitorica-Yrezabal, I.J.; Tuna, F.; Timco, G.A.; et al. A Modular Design of Molecular Qubits to Implement Universal Quantum Gates. *Nat. Commun.* **2016**, *7*, 11377. [CrossRef] [PubMed]
13. Atzori, M.; Chiesa, A.; Morra, E.; Chiesa, M.; Sorace, L.; Carretta, S.; Sessoli, R. A Two-Qubit Molecular Architecture for Electron-Mediated Nuclear Quantum Simulation. *Chem. Sci.* **2018**, *9*, 6183–6192. [CrossRef] [PubMed]
14. Borilovic, I.; Alonso, P.J.; Roubeau, O.; Aromí, G. A Bis-Vanadyl Coordination Complex as a 2-Qubit Quantum Gate. *Chem. Commun.* **2020**, *56*, 3139–3142. [CrossRef] [PubMed]

15. Troiani, F.; Affronte, M. Molecular Spins for Quantum Information Technologies. *Chem. Soc. Rev.* **2011**, *40*, 3119. [CrossRef]
16. Nakazawa, S.; Nishida, S.; Ise, T.; Yoshino, T.; Mori, N.; Rahimi, R.D.; Sato, K.; Morita, Y.; Toyota, K.; Shiomi, D.; et al. A Synthetic Two-Spin Quantum Bit: G -Engineered Exchange-Coupled Biradical Designed for Controlled-NOT Gate Operations. *Angew. Chem. Int. Ed.* **2012**, *51*, 9860–9864. [CrossRef]
17. Maniaki, D.; Garay-Ruiz, D.; Barrios, L.A.; Martins, D.O.T.A.; Aguilà, D.; Tuna, F.; Reta, D.; Roubeau, O.; Bo, C.; Aromí, G. Unparalleled Selectivity and Electronic Structure of Heterometallic [LnLn′Ln] Molecules as 3-Qubit Quantum Gates. *Chem. Sci.* **2022**, *13*, 5574–5581. [CrossRef]
18. Macaluso, E.; Rubín, M.; Aguilà, D.; Chiesa, A.; Barrios, L.A.; Martínez, J.I.; Alonso, P.J.; Roubeau, O.; Luis, F.; Aromí, G.; et al. A Heterometallic [LnLn′Ln] Lanthanide Complex as a Qubit with Embedded Quantum Error Correction. *Chem. Sci.* **2020**, *11*, 10337–10343. [CrossRef]
19. Aguilà, D.; Barrios, L.A.; Velasco, V.; Roubeau, O.; Repollés, A.; Alonso, P.J.; Sesé, J.; Teat, S.J.; Luis, F.; Aromí, G. Heterodimetallic [LnLn′] Lanthanide Complexes: Toward a Chemical Design of Two-Qubit Molecular Spin Quantum Gates. *J. Am. Chem. Soc.* **2014**, *136*, 14215–14222. [CrossRef]
20. Mata, J.A.; Hahn, F.E.; Peris, E. Heterometallic Complexes, Tandem Catalysis and Catalytic Cooperativity. *Chem. Sci.* **2014**, *5*, 1723–1732. [CrossRef]
21. Maldonado, C.S.; de la Rosa, J.R.; Lucio-Ortiz, C.J.; Sandoval-Rangel, L.; Martínez-Vargas, D.X.; Sánchez, R.A.L. Applications of Heterometallic Complexes in Catalysis. In *Direct Synthesis of Metal Complexes*; Elsevier: Amsterdam, The Netherlands, 2018; pp. 369–377.
22. Buchwalter, P.; Rosé, J.; Braunstein, P. Multimetallic Catalysis Based on Heterometallic Complexes and Clusters. *Chem. Rev.* **2015**, *115*, 28–126. [CrossRef]
23. Zhang, Y.-Y.; Gao, W.-X.; Lin, L.; Jin, G.-X. Recent Advances in the Construction and Applications of Heterometallic Macrocycles and Cages. *Coord. Chem. Rev.* **2017**, *344*, 323–344. [CrossRef]
24. Uber, J.S.; Estrader, M.; Garcia, J.; Lloyd-Williams, P.; Sadurní, A.; Dengler, D.; van Slageren, J.; Chilton, N.F.; Roubeau, O.; Teat, S.J.; et al. Molecules Designed to Contain Two Weakly Coupled Spins with a Photoswitchable Spacer. *Chem. Eur. J.* **2017**, *23*, 13648–13659. [CrossRef]
25. Wernsdorfer, W.; Mailly, D.; Timco, G.A.; Winpenny, R.E.P. Resonant Photon Absorption and Hole Burning in {Cr$_7$Ni} Antiferromagnetic Rings. *Phys. Rev. B* **2005**, *72*, 060409. [CrossRef]
26. Dey, A.; Acharya, J.; Chandrasekhar, V. Heterometallic 3d–4f Complexes as Single-Molecule Magnets. *Chem. Asian J.* **2019**, *14*, 4433–4453. [CrossRef]
27. Larsen, E.M.H.; Bonde, N.A.; Weihe, H.; Ollivier, J.; Vosch, T.; Lohmiller, T.; Holldack, K.; Schnegg, A.; Perfetti, M.; Bendix, J. Experimental Assignment of Long-Range Magnetic Communication through Pd & Pt Metallophilic Contacts. *Chem. Sci.* **2023**, *14*, 266–276. [CrossRef] [PubMed]
28. Funes, A.V.; Perfetti, M.; Kern, M.; Rußegger, N.; Carrella, L.; Rentschler, E.; Slageren, J.; Alborés, P. Single Molecule Magnet Features in the Butterfly [Co$^{III}_2$Ln$^{III}_2$] Pivalate Family with Alcohol-Amine Ligands. *Eur. J. Inorg. Chem.* **2021**, *2021*, 3191–3210. [CrossRef]
29. Belli Dell'Amico, D.; Ciattini, S.; Fioravanti, L.; Labella, L.; Marchetti, F.; Mattei, C.A.; Samaritani, S. The Heterotopic Divergent Ligand N-Oxide-4,4′-Bipyridine (BipyMO) as Directing-Agent in the Synthesis of Oligo- or Polynuclear Heterometallic Complexes. *Polyhedron* **2018**, *139*, 107–115. [CrossRef]
30. Sanz, S.; O'Connor, H.M.; Pineda, E.M.; Pedersen, K.S.; Nichol, G.S.; Mønsted, O.; Weihe, H.; Piligkos, S.; McInnes, E.J.L.; Lusby, P.J.; et al. [Cr$^{III}_8$M$^{II}_6$]$^{12+}$ Coordination Cubes (M^{II} = Cu, Co). *Angew. Chem. Int. Ed.* **2015**, *54*, 6761–6764. [CrossRef]
31. Wise, M.D.; Holstein, J.J.; Pattison, P.; Besnard, C.; Solari, E.; Scopelliti, R.; Bricogne, G.; Severin, K. Large, Heterometallic Coordination Cages Based on Ditopic Metallo-Ligands with 3-Pyridyl Donor Groups. *Chem. Sci.* **2015**, *6*, 1004–1010. [CrossRef]
32. Timco, G.A.; Carretta, S.; Troiani, F.; Tuna, F.; Pritchard, R.J.; Muryn, C.A.; McInnes, E.J.L.; Ghirri, A.; Candini, A.; Santini, P.; et al. Engineering the Coupling between Molecular Spin Qubits by Coordination Chemistry. *Nat. Nanotechnol.* **2009**, *4*, 173–178. [CrossRef]
33. Durot, S.; Taesch, J.; Heitz, V. Multiporphyrinic Cages: Architectures and Functions. *Chem. Rev.* **2014**, *114*, 8542–8578. [CrossRef]
34. Pilgrim, B.S.; Champness, N.R. Metal-Organic Frameworks and Metal-Organic Cages—A Perspective. *Chempluschem* **2020**, *85*, 1842–1856. [CrossRef] [PubMed]
35. O'Connor, H.M.; Sanz, S.; Pitak, M.B.; Coles, S.J.; Nichol, G.S.; Piligkos, S.; Lusby, P.J.; Brechin, E.K. [Cr$^{III}_8$M$^{II}_6$]$^{N+}$ (M^{II} = Cu, Co) Face-Centred, Metallosupramolecular Cubes. *CrystEngComm* **2016**, *18*, 4914–4920. [CrossRef]
36. O'Connor, H.M.; Sanz, S.; Scott, A.J.; Pitak, M.B.; Klooster, W.T.; Coles, S.J.; Chilton, N.F.; McInnes, E.J.L.; Lusby, P.J.; Weihe, H.; et al. [Cr$^{III}_8$M$^{II}_6$]$^{N+}$ Heterometallic Coordination Cubes. *Molecules* **2021**, *26*, 757. [CrossRef] [PubMed]
37. Sanz, S.; O'Connor, H.M.; Martí-Centelles, V.; Comar, P.; Pitak, M.B.; Coles, S.J.; Lorusso, G.; Palacios, E.; Evangelisti, M.; Baldansuren, A.; et al. [M$^{III}_2$M$_3$]$^{N+}$ Trigonal Bipyramidal Cages Based on Diamagnetic and Paramagnetic Metalloligands. *Chem. Sci.* **2017**, *8*, 5526–5535. [CrossRef] [PubMed]
38. Bruker. *Bruker APEX 2*; Bruker AXS Inc.: Madison, WI, USA, 2012.
39. Altomare, A.; Cuocci, C.; Giacovazzo, C.; Moliterni, A.; Rizzi, R.; Corriero, N.; Falcicchio, A. EXPO2013: A Kit of Tools for Phasing Crystal Structures from Powder Data. *J. Appl. Crystallogr.* **2013**, *46*, 1231–1235. [CrossRef]

40. Coelho, A.A. TOPAS and TOPAS-Academic: An Optimization Program Integrating Computer Algebra and Crystallographic Objects Written in C++. *J. Appl. Crystallogr.* **2018**, *51*, 210–218. [CrossRef]
41. Macrae, C.F.; Sovago, I.; Cottrell, S.J.; Galek, P.T.A.; McCabe, P.; Pidcock, E.; Platings, M.; Shields, G.P.; Stevens, J.S.; Towler, M.; et al. Mercury 4.0: From Visualization to Analysis, Design and Prediction. *J. Appl. Crystallogr.* **2020**, *53*, 226–235. [CrossRef]
42. Bain, G.A.; Berry, J.F. Diamagnetic Corrections and Pascal's Constants. *J. Chem. Educ.* **2008**, *85*, 532. [CrossRef]
43. Glidewell, C. Metal Acetylacetonate Complexes: Preparation and Characterization. In *Inorganic Experiments*; Woollins, J.E., Ed.; Wiley-VCH: Weinheim, Germany, 2003.
44. Faggi, E.; Gavara, R.; Bolte, M.; Fajarí, L.; Juliá, L.; Rodríguez, L.; Alfonso, I. Copper(Ii) Complexes of Macrocyclic and Open-Chain Pseudopeptidic Ligands: Synthesis, Characterization and Interaction with Dicarboxylates. *Dalton Trans.* **2015**, *44*, 12700–12710. [CrossRef]
45. Flores-Rojas, G.G.; Ruiu, A.; Vonlanthen, M.; Rojas-Montoya, S.M.; Martínez-Serrano, R.D.; Morales-Morales, D.; Rivera, E. Synthesis and Characterization of Cyclen Cored Photoactive Star Compounds and Their Cu(I) and Cu(II) Complexes. Effect of the Valence and Ligand Size on Their Molar Extinction Coefficient. *Inorg. Chim. Acta* **2020**, *513*, 119927. [CrossRef]
46. Tosato, M.; Dalla Tiezza, M.; May, N.V.; Isse, A.A.; Nardella, S.; Orian, L.; Verona, M.; Vaccarin, C.; Alker, A.; Mäcke, H.; et al. Copper Coordination Chemistry of Sulfur Pendant Cyclen Derivatives: An Attempt to Hinder the Reductive-Induced Demetalation in 64/67 Cu Radiopharmaceuticals. *Inorg. Chem.* **2021**, *60*, 11530–11547. [CrossRef] [PubMed]
47. El Ghachtouli, S.; Cadiou, C.; Déchamps-Olivier, I.; Chuburu, F.; Aplincourt, M.; Roisnel, T. (Cyclen– and Cyclam–Pyridine)Copper Complexes: The Role of the Pyridine Moiety in CuII and CuI Stabilisation. *Eur. J. Inorg. Chem.* **2006**, *2006*, 3472–3481. [CrossRef]
48. Lacerda, S.; Campello, M.P.; Santos, I.C.; Santos, I.; Delgado, R. Study of the Cyclen Derivative 2-[1,4,7,10-Tetraazacyclododecan-1-Yl]-Ethanethiol and Its Complexation Behaviour towards d-Transition Metal Ions. *Polyhedron* **2007**, *26*, 3763–3773. [CrossRef]
49. Stoll, S.; Schweiger, A. EasySpin, a Comprehensive Software Package for Spectral Simulation and Analysis in EPR. *J. Magn. Reson.* **2006**, *178*, 42–55. [CrossRef] [PubMed]
50. Bonomo, R.P.; Di Bilio, A.J.; Riggi, F. EPR Investigation of Chromium(III) Complexes: Analysis of Their Frozen Solution and Magnetically Dilute Powder Spectra. *Chem. Phys.* **1991**, *151*, 323–333. [CrossRef]
51. Elbers, G.; Remme, S.; Lehmann, G. EPR of Chromium(3+) in Tris(Acetylacetonato)Gallium(III) Single Crystals. *Inorg. Chem.* **1986**, *25*, 896–897. [CrossRef]
52. Jeschke, G. DEER Distance Measurements on Proteins. *Annu. Rev. Phys. Chem.* **2012**, *63*, 419–446. [CrossRef]
53. Chilton, N.F.; Anderson, R.P.; Turner, L.D.; Soncini, A.; Murray, K.S. PHI: A Powerful New Program for the Analysis of Anisotropic Monomeric and Exchange-Coupled Polynuclear d- and f -Block Complexes. *J. Comput. Chem.* **2013**, *34*, 1164–1175. [CrossRef]

Disclaimer/Publisher's Note: The statements, opinions and data contained in all publications are solely those of the individual author(s) and contributor(s) and not of MDPI and/or the editor(s). MDPI and/or the editor(s) disclaim responsibility for any injury to people or property resulting from any ideas, methods, instructions or products referred to in the content.

Article

Synthesis, Structure, and Characterizations of a Heterobimetallic Heptanuclear Complex [Pb$_2$Co$_5$(acac)$_{14}$]

Yuxuan Zhang, Zheng Wei and Evgeny V. Dikarev *

Department of Chemistry, University at Albany, State University of New York, 1400 Washington Ave, Albany, NY 12222, USA; yzhang50@albany.edu (Y.Z.); zwei@albany.edu (Z.W.)
* Correspondence: edikarev@albany.edu

Abstract: An unusual heterobimetallic volatile compound [Pb$_2$Co$_5$(acac)$_{14}$] was synthesized by the gas phase/solid-state technique. The preparation can be readily scaled up using the solution approach. X-ray powder diffraction, ICP-OES analysis, and DART mass spectrometry were engaged to confirm the composition and purity of heterobimetallic complex. The composition is unique among the large family of lead(tin): transition metal = 2:1, 1:1, and 1:2 β-diketonates compounds that are mostly represented by coordination polymers. The molecular structure of the complex was elucidated by synchrotron single crystal X-ray diffraction to reveal the unique heptanuclear moiety {Co(acac)$_2$[Pb(acac)$_2$-Co(acac)$_2$-Co(acac)$_2$]$_2$} built upon bridging interactions of acetylacetonate oxygens to neighboring metal centers that bring their coordination numbers to six. The appearance of unique heptanuclear assembly can be attributed to the fact that the [Co(acac)$_2$] units feature both *cis*- and *trans-bis*-bridging modes, making the polynuclear moiety rather flexible. This type of octahedral coordination is relatively unique among known lead(tin)-3*d* transition metal β-diketonates. Due to the high-volatility, [Pb$_2$Co$_5$(acac)$_{14}$] can be potentially applied as a MOCVD precursor for the low-temperature preparation of lead-containing functional materials.

Keywords: single crystal; lead complexes; cobalt complexes; heterometallic structures; heptanuclear molecule

---

Citation: Zhang, Y.; Wei, Z.; Dikarev, E.V. Synthesis, Structure, and Characterizations of a Heterobimetallic Heptanuclear Complex [Pb$_2$Co$_5$(acac)$_{14}$]. *Crystals* **2023**, *13*, 1089. https://doi.org/10.3390/cryst13071089

Academic Editors: Antonio Bianchi and Matteo Savastano

Received: 8 June 2023
Revised: 25 June 2023
Accepted: 10 July 2023
Published: 12 July 2023

Copyright: © 2023 by the authors. Licensee MDPI, Basel, Switzerland. This article is an open access article distributed under the terms and conditions of the Creative Commons Attribution (CC BY) license (https://creativecommons.org/licenses/by/4.0/).

1. Introduction

A number of divalent lead-transition metal heterobimetallic compounds Pb$_x$M$_y$(β-dik)$_{2x+2y}$ (M = Mn, Co, Ni, Fe, Zn) have been synthesized as single-source precursors (SSPs) for functional lead-containing transition-metal oxides or fluorides [1–3]. β-Diketonate ligands such as acetylacetonate (*acac*) and fluorinated hexafluoroacetylacetonate (*hfac*) and trifluoroacetylacetonate (*tfac*) have been employed to form heterometallic complexes that exhibit good volatility and solubility. More importantly, the ligands endowed the above SSPs with low-temperature decomposition to produce target materials such as magnetoelectrics Pb$_2$MF$_6$ (M = Ni and Co) [1,4] or multiferroics Pb$_2$Fe$_2$O$_5$ [3,5]. Most of the lead-transition metal β-diketonates appeared to have polymeric structures built upon combination of Pb(β-dik)$_y$ (y = 0, 1, 2) and M(β-dik)$_x$ (x = 2, 3) units connected by Lewis acid–base Pb–O/M–O interactions (bridging bonds) with three different Pb:M ratios of 2:1, 1:1, and 1:2 [1–3]. Within the heterometallic assemblies, lead and transition metals both exhibit coordination numbers of six with the former preferring *cis*-bridging and the latter *trans*-bridging with neighboring fragments [1–3,6–14]. In this work, we describe the synthesis and characterization of a new lead-containing heterobimetallic heptanuclear homoleptic complex [Pb$_2$Co$_5$(acac)$_{14}$] which is distinguished from the previously reported Pb–M compounds in several aspects. The ratio of Pb:Co = 2:5 is unique, while the heptanuclear structure is constructed by both *cis*- and *trans*-bridged [Co(acac)$_2$] units providing more complex/flexible coordination environment. Due to the good volatility, the [Pb$_2$Co$_5$(acac)$_{14}$] complex is interesting as a prospective precursor in the metal–organic chemical vapor deposition (MOCVD) for the preparation of high-technological thin films with desired stoichiometry [15–17]. The unusual lead:cobalt ratio makes it attractive to investigate the potential application for the low-temperature formation of lead–cobalt-coated anodes, lone-pair multiferroics, and superconducting materials [18–27].

2. Materials and Methods

2.1. Materials and Measurements

Lead(II) acetylacetonate [Pb(acac)$_2$] and cobalt(II) acetylacetonate [Co(acac)$_2$] were purchased from Sigma-Aldrich and used as received after checking their powder X-ray diffraction patterns. The ICP-OES analysis was carried out on ICPE-9820 plasma atomic emission spectrometer, Shimadzu. The IR spectrum was measured using IRTracer-100 Fourier Transform Infrared Spectrophotometer, Shimadzu. The DART-MS spectra were recorded on AccuTof 4G LC-plus DART mass spectrometer, JEOL. X-ray powder diffraction data were collected on a Rigaku MiniFlex 6G benchtop diffractometer (Cu Kα radiation, D/teX Ultra silicon strip one-dimensional detector, step of 0.01° 2θ, 20 °C). Le Bail fit refinement for powder diffraction patterns has been performed using TOPAS, version 4 software package (Bruker AXS, Billerica, MA, USA, 2006).

2.2. General Synthetic Procedures

The solid-state/gas phase synthesis of heterometallic complexes was carried out by grinding and sealing the stoichiometric mixture of starting reagents in an evacuated glass ampule and placing it into a gradient furnace. Solution synthesis of heterometallic complex was performed by adding dry and deoxygenated organic solvent to the stoichiometric mixture of starting reagents in a flask under a dry, oxygen-free argon atmosphere using standard Schlenk and glove box techniques. Detailed synthetic procedures are described in Section 3.1.

2.3. X-ray Crystallographic Procedures

The crystals of [Pb$_2$Co$_5$(acac)$_{14}$] were immersed in cryo-oil, mounted on a glass fiber, and measured at the temperature of 100(2) K. The X-ray diffraction data were collected on a Huber Kappa system with a DECTRIS PILATUS3 X 2M(CdTe) pixel array detector using ϕ scans (synchrotron radiation at λ = 0.49594 Å) located at the Advanced Photon Source, Argonne National Laboratory (NSF's ChemMatCARS, Sector 15, Beamline 15-ID-D). The crystals of [PbCo(acac)$_4$] were immersed in cryo-oil, mounted on a loop, and measured at the temperature of 100(2) K. The X-ray diffraction data were collected on a Bruker D8 SMART diffractometer using Mo Kα radiation. The dataset reduction and integration for both structures were performed with the Bruker software package SAINT (version 8.38A) [28]. The data were corrected for absorption effects using the empirical methods as implemented in SADABS (version 2016/2) [29]. The structures were solved by SHELXT (version 2018/2) [30] and refined by full-matrix least-squares procedures using the Bruker SHELXTL (version 2019/2) [31] software package through the OLEX2 graphical interface [32]. All non-hydrogen atoms were refined anisotropically. Hydrogen atoms were included in idealized positions for the structure factor calculations with U_{iso}(H) = 1.2 U_{eq}(C) and U_{iso}(H) = 1.5 U_{eq}(C) for methyl groups. Crystallographic data and details of the data collection and structure refinement are listed in Table 1.

Table 1. Crystal data and structure refinement parameters for [Pb$_2$Co$_5$(acac)$_{14}$] and [PbCo(acac)$_4$].

Compound	Pb$_2$Co$_5$(acac)$_{14}$	PbCo(acac)$_4$
CCDC	2268578	2268579
Empirical formula	C$_{70}$H$_{98}$Co$_5$Pb$_2$O$_{28}$	C$_{20}$H$_{28}$CoPbO$_8$
Formula weight	2096.51	662.54
Temperature (K)	100(2)	100(2)
Wavelength (Å)	0.49594	0.71073
Crystal system	Monoclinic	Monoclinic
Space group	$P2_1/c$	$P2_1/c$
a (Å)	23.8452(6)	8.7792(15)
b (Å)	10.7063(3)	20.142(3)
c (Å)	16.3561(4)	13.720(2)
β (°)	97.0960(10)	91.957(2)
V (Å^3)	4143.63(19)	2424.8(7)
Z	2	4
ρ_{calcd} (g·cm^{-3})	1.680	1.815
μ (mm^{-1})	1.989	7.657
F(000)	2082	1284
Crystal size (mm^3)	0.09 × 0.07 × 0.04	0.16 × 0.13 × 0.10
θ range for data collection (°)	1.201–19.317	2.321–28.277
Reflections collected	129,052	20,649
Independent reflections	10,257 [R_{int} = 0.0596]	5640 [R_{int} = 0.0783]
Transmission factors (min/max)	0.7421/0.8414	0.6241/0.7563
Completeness to full θ (%)	99.6	99.7
Data/restraints/params.	10,257/0/490	5640/0/279
R1,[a] wR2[b] ($I > 2\sigma(I)$)	0.0198, 0.0533	0.0545/0.1296
R1,[a] wR2[b] (all data)	0.0225, 0.0542	0.0975/0.1496
Quality-of-fit[c]	1.064	1.029

[a] $R1 = \Sigma||F_o| - |F_c||/\Sigma|F_o|$. [b] $wR2 = [\Sigma[w(F_o^2 - F_c^2)^2]/\Sigma[w(F_o^2)^2]]^{\frac{1}{2}}$. [c] Quality-of-fit = $[\Sigma[w(F_o^2 - F_c^2)^2]/(N_{obs} - N_{params})]^{\frac{1}{2}}$, based on all data.

3. Results and Discussion

3.1. Synthesis and Properties of [Pb$_2$Co$_5$(acac)$_{14}$] and [PbCo(acac)$_4$]

Synthesis of heterometallic complex [Pb$_2$Co$_5$(acac)$_{14}$] was carried out by both solid-state and solution approaches through the stoichiometric reaction (Equation (1)). For the solid-state synthesis, 10 mg of [Pb(acac)$_2$] (0.025 mmol) and 16 mg of [Co(acac)$_2$] (0.062 mmol) were sealed in an evacuated glass ampule. The ampule was placed in a furnace with a temperature gradient of 130 to 120 °C. After one day of heating, block-shaped violet [Pb$_2$Co$_5$(acac)$_{14}$] crystals were sublimed to the cold zone of the container (16 mg, ca. 62% yield). The product can be obtained in nearly quantitative yield by extending the reaction time to one week under the same conditions. Microcrystalline powder of [Pb$_2$Co$_5$(acac)$_{14}$] was synthesized by dissolving 100 mg of [Pb(acac)$_2$] (0.25 mmol) and 160 mg of [Co(acac)$_2$] (0.62 mmol) in 20 mL of dry and deoxygenated hexanes under argon atmosphere. The solution was stirred at room temperature for 1 day, resulting in a large amount of violet precipitate being formed. The solid was filtered off and dried under vacuum at 100 °C sand bath overnight to afford the final product with a practically quantitative yield.

$$2\text{Pb(acac)}_2 + 5\text{Co(acac)}_2 \rightarrow \text{Pb}_2\text{Co}_5(\text{acac})_{14} \qquad (1)$$

The block-shaped violet crystals acquired from the solid-state reaction were first checked by the ICP-OES analysis, which revealed the metal ratio of Pb:Co as 2:5, different from any previously investigated Pb$_x$M$_y$(β-dik)$_{2x+2y}$ complexes [1–3]. X-ray powder diffraction was applied to check the phase purity of the [Pb$_2$Co$_5$(acac)$_{14}$] bulk products obtained from the solid state and solution methods (Figure 1). The Le Bail fit was performed to confirm that the experimental powder patterns of the bulk products correspond to the theoretical spectrum calculated from the single crystal X-ray data (Figure 1 and Table 2).

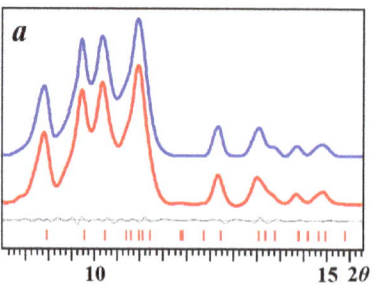 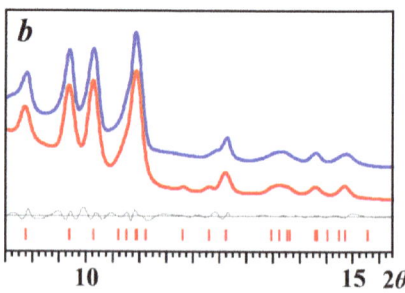

Figure 1. X-ray powder diffraction patterns of (**a**) [Pb$_2$Co$_5$(acac)$_{14}$] bulk product obtained from the solution reaction and the Le Bail fit; (**b**) [Pb$_2$Co$_5$(acac)$_{14}$] crystals obtained from the solid-state reaction and the Le Bail fit. In both pictures, blue and red curves represent experimental and calculated patterns, respectively. Gray is the differential curve with theoretical peak positions shown as red bars at the bottom.

Table 2. Comparison of the unit cell parameters for [Pb$_2$Co$_5$(acac)$_{14}$] product obtained from the single crystal refinement and the Le Bail fit.

	Single Crystal Data (−173 °C)	Solution Product Le Bail Fit (20 °C)	Solid-State Product Le Bail Fit (20 °C)
Space group	$P2_1/c$	$P2_1/c$	$P2_1/c$
a (Å)	23.8452(6)	23.939(3)	23.933(8)
b (Å)	10.7063(3)	10.9156(10)	10.908(4)
c (Å)	16.3561(4)	16.399(2)	16.496(5)
β (°)	97.0960(10)	96.767(11)	97.026(13)
V (Å^3)	4143.63(19)	4255.5(9)	4274(2)

Direct Analysis in Real Time (DART) mass spectrometry was performed to check the retention of heterometallic structure and analyze the ions in the gas phase. In the positive mode mass spectrum of [Pb$_2$Co$_5$(acac)$_{14}$] (SI, Figure S1 and Table S1) heterometallic ion peaks such as [PbCo$_3$(acac)$_7$]$^+$, [PbCo$_4$(acac)$_9$]$^+$, and [Pb$_2$Co(acac)$_5$]$^+$ that represent the fragmentation of the heptanuclear [Pb$_2$Co$_5$(acac)$_{14}$] molecule can be clearly detected. All heterometallic ion fragments with relatively high intensities are measured precisely with the difference between measured and calculated m/z being smaller than 0.006 and characteristic isotope distribution patterns (SI, Figure S1 and Table S1). The peak corresponding to heptanuclear ion [M-L]$^+$ (M = [Pb$_2$Co$_5$(acac)$_{14}$], L = $acac$) was not detected in the DART mass spectrum, indicating that the structure is quite fragile. In fact, this is the first heterometallic molecule in our research that was found not to feature the [M-L]$^+$ peak in the positive mode, though we should notice that it is the heaviest so far to be investigated. At the same time, the fragment peaks unambiguously confirm the (+2) oxidation states of both lead and cobalt.

The IR spectrum of [Pb$_2$Co$_5$(acac)$_{14}$] was also recorded and shown in the SI, Figure S2. The [Pb$_2$Co$_5$(acac)$_{14}$] complex is stable in the presence of oxygen in the solid state, but is quite sensitive to moisture. It has a good solubility in polar, weakly/non-coordinating solvents such as CHCl$_3$ and CH$_2$Cl$_2$ at room temperature, but shows a poor solubility in non-polar solvents such as hexanes, pentanes, cyclohexane, and toluene at room temperature. [Pb$_2$Co$_5$(acac)$_{14}$] can be quantitatively sublimed at as low as 105 °C under static vacuum conditions (sealed evacuated ampule) and starts to show the traces of decomposition when the temperature is raised to 150 °C.

The synthesis of heterometallic complex [PbCo(acac)$_4$] was carried out by the solid-state approach. First, 15 mg of [Pb(acac)$_2$] (0.037 mmol) and 9 mg of [Co(acac)$_2$] (0.035 mmol) were sealed in an evacuated glass ampule. The ampule was placed in a furnace with a temperature gradient of 115 to 105 °C. After one day of heating, block-shaped pink crystals of [PbCo(acac)$_4$] were sublimed to the "cold" zone of the container (10 mg, ca. 42% yield). The differences in synthetic procedures for the preparation of [Pb$_2$Co$_5$(acac)$_{14}$] and [PbCo(acac)$_4$] are the ratio of starting reagents and the temperature in the "hot" zone of the ampule. The crystal colors of two compounds are noticeably different. [PbCo(acac)$_4$] is structurally analogous to the previously reported [PbM(β-dik)$_4$] compounds (M = Mn, Fe, and Zn; β-dik = *acac*, *tfac*, and *hfac*) [1–3] and was obtained for comparison of bond distances and coordination mode of Co ion, which are discussed in detail in Section 3.2.

3.2. Single Crystal Structure of [Pb$_2$Co$_5$(acac)$_{14}$]

In this work, we used sterically uncongested β-diketonate ligand, acetylacetonate (*acac*), to synthesize a hetero*bi*metallic assembly [Pb$_2$Co$_5$(acac)$_{14}$], which contains both *cis*- and *trans*-bridging [Co(acac)$_2$] units. Single-crystal X-ray diffraction analysis revealed that the solid-state structure of this heterometallic complex contains discrete heptanuclear molecules with metal chain in an order of [Co-Co-Pb-Co-Pb-Co-Co]. The central Co1 atom (Figure 2) sits on an inversion center; therefore, there are only four crystallographically independent metal sites (three Co and one Pb) within the structural motif. Each metal center has two chelating *acac* ligands and fulfills its coordination environment by two additional contacts with oxygen atoms that are chelating its neighbors. The [M(acac)$_2$] (M = Co, Pb) groups are connected through Lewis acid–base M–O interactions of 2.09–2.26 and 2.81–2.86 Å for Co and Pb ions, respectively (Tables 3 and 4), that are significantly shorter than the sum of the corresponding van der Waals radii. Although all metal centers have a coordination number of 6, the coordination environments are distinctively different. Two outmost [Co(acac)$_2$] fragments offer only one *acac* oxygen each for the bridging interactions with the neighboring [Co(acac)$_2$] units, which, in turn, offer two *acac* oxygens for bridging interactions with the end-chain [Co(acac)$_2$] units in a *cis*-fashion and an additional *acac* oxygen to bridge the [Pb(acac)$_2$] unit. The bond distances for bridging Co–O bonds in these units are between 2.09 and 2.22 Å, longer than chelating and chelating–bridging Co–O distances in the structural motif (Table 3). The cobalt atoms in these two units maintain a slightly distorted octahedral geometry with two *acac* ligand planes being almost perpendicular to each other. Since both cobalt centers in the [Co(acac)$_2$] units appear as *bis-cis*-chelated, they are obviously chiral. The Co2 and Co3 exhibit the same chirality (either Δ,Δ or Λ,Λ). Considering the inversion center in the middle of this heptanuclear assembly, the compound is *meso* and does not feature diastereomers. In the central [Co(acac)$_2$] fragment, Co1 ion has two chelating *acac* ligands that are located in a plane, with two bridging Co–O interactions from the neighboring [Pb(acac)$_2$] units in a *trans*-configuration. Both *acac* oxygens from one of the ligand in the central [Co(acac)$_2$] group are pure chelating, while those from the second ligand bridge to [Pb(acac)$_2$] groups on both sides. Since these two *trans*-bridging Co–O distances are much longer than those four from chelating bonds, the coordination of the central Co ion can be described as an axially elongated octahedral geometry. In the heptanuclear structure, the [Pb(acac)$_2$] units act as a glue to connect two [Co$_2$(acac)$_4$] fragments on both sides with the central [Co(acac)$_2$] unit by two bridging interactions through one oxygen from each of two *acac* ligands. The $6s^2$ lone electron pair of the lead repels two Pb-chelating *acac* ligands to be in a face-to-face fashion arrangement with a dihedral angle of 33.01°.

The appearance of both *cis*- and *trans*-bridging modes for the [Co(acac)$_2$] units in [Pb$_2$Co$_5$(acac)$_{14}$] makes it unique among polynuclear cobalt and heterometallic lead–cobalt β-diketonate complexes. Thus, the tetranuclear structure of [Co$_4$(acac)$_8$] (Figure 3) features only the *cis*-bridging mode to combine four [Co(acac)$_2$] monomers together [6]. In contrast, the polymeric heterometallic structure of [PbCo(acac)$_4$] (Figure 4) shows only *trans*-bridged [Co(acac)$_2$] building up the 1D polymeric chain with [Pb(acac)$_2$] units in 1:1 ratio. Analogous β-diketonate [PbCo(hfac)$_4$] [1] also has a polymeric structure but it is constructed by alternating [Co(hfac)$_3$]$^-$ and [Pb(acac)]$^+$ units. In turn, all [Pb(acac)$_2$] fragments in the lead-transition metal β-diketonate structures [1–3] are *cis*-bridging.

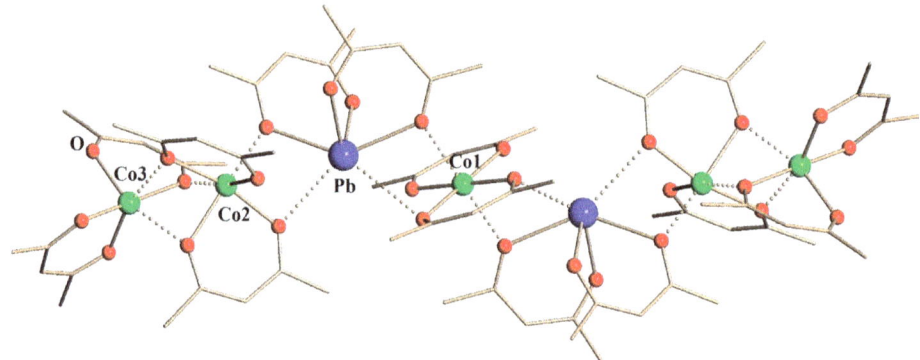

Figure 2. Molecular structure of [Pb$_2$Co$_5$(acac)$_{14}$] complex. Hydrogen atoms are omitted for clarity. All bridging M–O bonds are shown as dotted lines. Full view of molecular structure can be found in the SI, Figure S3.

Table 3. Co–O bond distances (Å) in the structures of [Pb$_2$Co$_5$(acac)$_{14}$] and in other cobalt-containing polynuclear/heterometallic complexes.

	Co–O$_c$	Co–O$_{c-b}$	Co–O$_b$ (cis)	Co–O$_b$ (trans)
Pb$_2$Co$_5$(acac)$_{14}$	1.98–2.05	2.02–2.08	2.09–2.22	2.26
Co$_4$(acac)$_8$	2.00–2.03	2.02–2.11	2.09–2.20	
PbCo(acac)$_4$	2.02–2.03	2.03–2.04		2.22–2.24
PbCo(hfac)$_4$	2.05	2.04–2.07		

c—chelating; c-b—chelating-bridging; b—bridging.

Table 4. Pb–O bond distances (Å) in the structures of [Pb$_2$Co$_5$(acac)$_{14}$] and other lead–cobalt β-diketonate complexes.

	Pb–O$_c$	Pb–O$_{c-b}$	Pb–O$_b$ (cis)
Pb$_2$Co$_5$(acac)$_{14}$	2.34–2.35	2.48–2.49	2.81–2.86
PbCo(acac)$_4$	2.31–2.32	2.51–2.52	2.86–2.89
PbCo(hfac)$_4$	2.32		2.74–2.81

c—chelating; c-b—chelating-bridging; b—bridging.

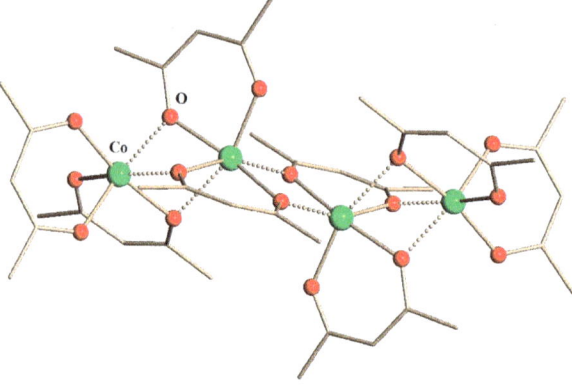

Figure 3. Molecular structure of tetranuclear [Co$_4$(acac)$_8$]. Hydrogen atoms are omitted for clarity. All bridging Co–O bonds are shown as dotted lines.

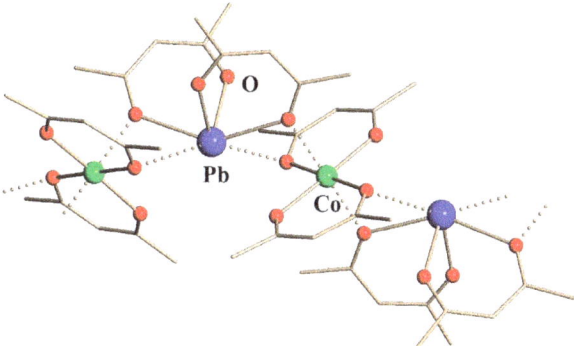

Figure 4. Fragment of crystal structure of polymeric [PbCo(acac)$_4$] compound. Hydrogen atoms are omitted for clarity. All bridging M–O bonds are shown as dotted lines. Full view of the structure can be found in the SI, Figure S4.

4. Conclusions

The hetero*bi*metallic complex [Pb$_2$Co$_5$(acac)$_{14}$] was synthesized by both solid-state and solution techniques and its unique heptanuclear molecular structure was revealed by synchrotron single-crystal X-ray diffraction. It was shown that the [Co(acac)$_2$] units in the assembly exhibit both *cis*- and *trans*-bridging modes, which is different from previously reported polynuclear cobalt complexes and lead–transition metal β-diketonate coordination polymers. With high volatility and solubility, this heterometallic compound with unprecedented transition metal:lead ratio can serve as a precursor for the low-temperature preparation of prospective functional thin film materials such as lead–cobalt composite coating, lone-pair multiferroics, and superconductors. Importantly, this unique molecule with two distinctively different cobalt positions should be investigated as a model structure for the design of more complex hetero*tri*metallic compounds with partial substitution of cobalt with other divalent metals.

Supplementary Materials: The following supporting information can be downloaded at: https://www.mdpi.com/article/10.3390/cryst13071089/s1, Figure S1: DART-mass spectrum of [Pb$_2$Co$_5$(acac)$_{14}$] in a positive mode; Table S1: Assignment of ions detected in a positive-ion DART-mass spectrum of [Pb$_2$Co$_5$(acac)$_{14}$]; Figure S2: IR spectrum of [Pb$_2$Co$_5$(acac)$_{14}$]; Figure S3: Crystal structure of [Pb$_2$Co$_5$(acac)$_{14}$]; Table S2: Bond distances (Å) and angles (°) in the structure of [Pb$_2$Co$_5$(acac)$_{14}$]; Figure S4: Crystal structure of PbCo(acac)$_4$; Table S3: Bond distances (Å) and angles (°) in the structure of [PbCo(acac)$_4$].

Author Contributions: Y.Z. performed synthesis, obtained suitable single crystals for X-ray diffraction analysis, carried out characterization, and prepared the draft of the manuscript; Z.W. performed synchrotron single-crystal measurement and refinement, and provided structural description; E.V.D. conceived and supervised the project and finalized the manuscript. All authors have read and agreed to the published version of the manuscript.

Funding: This work was supported by the National Science Foundation (CHE-1955585).

Data Availability Statement: Data in this study are available upon request.

Acknowledgments: NSF's ChemMatCARS, Sector 15 at the Advanced Photon Source (APS), Argonne National Laboratory (ANL) is supported by the Divisions of Chemistry (CHE) and Materials Research (DMR), National Science Foundation, under grant number NSF/CHE-1834750. This research used resources of the Advanced Photon Source, a US Department of Energy (DOE) Office of Science user facility operated for the DOE Office of Science by Argonne National Laboratory under Contract No. DE-AC02-06CH11357.

Conflicts of Interest: The authors declare no conflict of interest.

References

1. Navulla, A.; Tsirlin, A.A.; Abakumov, A.M.; Shpanchenko, R.V.; Zhang, H.; Dikarev, E.V. Fluorinated Heterometallic β-Diketonates as Volatile Single-Source Precursors for the Synthesis of Low-Valent Mixed-Metal Fluorides. *J. Am. Chem. Soc.* **2011**, *133*, 692–694. [CrossRef] [PubMed]
2. Zhang, H.; Yang, J.-H.; Shpanchenko, R.V.; Abakumov, A.M.; Hadermann, J.; Clérac, R.; Dikarev, E.V. New Class of Single-Source Precursors for the Synthesis of Main Group-Transition Metal Oxides: Heterobimetallic Pb-Mn Beta-Diketonates. *Inorg. Chem.* **2009**, *48*, 8480–8488. [CrossRef]
3. Lieberman, C.M.; Navulla, A.; Zhang, H.; Filatov, A.S.; Dikarev, E.V. Mixed-Ligand Approach to Design of Heterometallic Single-Source Precursors with Discrete Molecular Structure. *Inorg. Chem.* **2014**, *53*, 4733–4738. [CrossRef] [PubMed]
4. Chrétien, A.; Samouël, M. Magnetische Eigenschaften der Verbindungen $BaMF_4$ und Pb_2MF_6 (M = Mn, Fe, Co, Ni, Cu, Zn). *Monatshefte Für Chemie.* **1972**, *103*, 17–23. [CrossRef]
5. Grenier, J.; Pouchard, M.; Hagenmuller, P. Caractérisations physiques du ferrites de plomb $Pb_2Fe_2O_5$. *Rev. Chim. Miner.* **1977**, *14*, 515–522.
6. Vreshch, V.D.; Yang, J.-H.; Zhang, H.; Filatov, A.S.; Dikarev, E.V. Monomeric Square-Planar Cobalt(II) Acetylacetonate: Mystery or Mistake? *Inorg. Chem.* **2010**, *49*, 8430–8434. [CrossRef]
7. Krisyuk, V.V.; Urkasym Kyzy, S.; Rybalova, T.V.; Baidina, I.A.; Korolkov, I.V.; Chizhov, D.L.; Bazhin, D.N.; Kudyakova, Y.S. Isomerization as a Tool to Design Volatile Heterometallic Complexes with Methoxy-Substituted β-Diketonates. *J. Coord. Chem.* **2018**, *71*, 2194–2208. [CrossRef]
8. Krisyuk, V.V.; Baidina, I.A.; Basova, T.V.; Bulusheva, L.G.; Igumenov, I.K. Self-assembly of Coordination Polymers from Volatile Pd II and Pb II B-diketonate Derivatives through Metallophilic Interactions: Self-Assembly of Coordination Polymers from PdIIand Pb II β-Diketonates. *Eur. J. Inorg. Chem.* **2013**, *2013*, 5738–5745. [CrossRef]
9. Krisyuk, V.V.; Baidina, I.A.; Kryuchkova, N.A.; Logvinenko, V.A.; Plyusnin, P.E.; Korolkov, I.V.; Zharkova, G.I.; Turgambaeva, A.E.; Igumenov, I.K. Volatile Heterometallics: Structural Diversity of Pd–Pb β-Diketonates and Correlation with Thermal Properties. *Dalton Trans.* **2017**, *46*, 12245–12256. [CrossRef]
10. Lieberman, C.M.; Barry, M.C.; Wei, Z.; Rogachev, A.Y.; Wang, X.; Liu, J.-L.; Clérac, R.; Chen, Y.-S.; Filatov, A.S.; Dikarev, E.V. Position Assignment and Oxidation State Recognition of Fe and Co Centers in Heterometallic Mixed-Valent Molecular Precursors for the Low-Temperature Preparation of Target Spinel Oxide Materials. *Inorg. Chem.* **2017**, *56*, 9574–9584. [CrossRef]
11. Maryunina, K.; Fokin, S.; Ovcharenko, V.; Romanenko, G.; Ikorskii, V. Solid Solutions: An Efficient Way to Control the Temperature of Spin Transition in Heterospin Crystals $M_xCu_{1-x}(Hfac)_2L$ (M = Mn, Ni, Co; L = nitronyl Nitroxide). *Polyhedron* **2005**, *24*, 2094–2101. [CrossRef]
12. Reis, S.G.; Del Águila-Sánchez, M.A.; Guedes, G.P.; Ferreira, G.B.; Novak, M.A.; Speziali, N.L.; López-Ortiz, F.; Vaz, M.G.F. Synthesis, Crystal Structures and Magnetic Behaviour of Four Coordination Compounds Constructed with a Phosphinic Amide-TEMPO Radical and $[M(Hfac)_2]$ (M = Cu(II), Co(II) and Mn(II)). *Dalton Trans.* **2014**, *43*, 14889–14901. [CrossRef] [PubMed]
13. Maxim, C.; Matni, A.; Geoffroy, M.; Andruh, M.; Hearns, N.G.R.; Clérac, R.; Avarvari, N. C3 Symmetric Tris(Phosphonate)-1,3,5-Triazine Ligand: Homopolymetallic Complexes and Its Radical Anion. *New J. Chem.* **2010**, *34*, 2319. [CrossRef]
14. Dikarev, E.V.; Zhang, H.; Li, B. Heterometallic Bismuth-Transition Metal Homoleptic Beta-Diketonates. *J. Am. Chem. Soc.* **2005**, *127*, 6156–6157. [CrossRef]
15. Hubert-Pfalzgraf, L.G. Some Trends in the Design of Homo- and Heterometallic Molecular Precursors of High-Tech Oxides. *Inorg. Chem. Commun.* **2003**, *6*, 102–120. [CrossRef]
16. Jones, A.C.; Aspinall, H.C.; Chalker, P.R. Molecular Design of Improved Precursors for the MOCVD of Oxides Used in Microelectronics. *Surf. Coat. Technol.* **2007**, *201*, 9046–9054. [CrossRef]
17. Weiss, F.; Audier, M.; Bartasyte, A.; Bellet, D.; Girardot, C.; Jimenez, C.; Kreisel, J.; Pignard, S.; Salaun, M.; Ternon, C. Multifunctional Oxide Nanostructures by Metal-Organic Chemical Vapor Deposition (MOCVD). *Pure Appl. Chem.* **2009**, *81*, 1523–1534. [CrossRef]
18. Nikoloski, A.N.; Nicol, M.J. Addition of Cobalt to Lead Anodes Used for Oxygen Evolution—A Literature Review. *Miner. Process. Extr. Met. Rev.* **2009**, *31*, 30–57. [CrossRef]
19. Felder, A.; Prengaman, R.D. Lead Alloys for Permanent Anodes in the Nonferrous Metals Industry. *Jom* **2006**, *58*, 28–31. [CrossRef]
20. Nikoloski, A.N.; Barmi, M.J. Novel Lead–Cobalt Composite Anodes for Copper Electrowinning. *Hydrometallurgy* **2013**, *137*, 45–52. [CrossRef]
21. Ivanov, I.; Stefanov, Y.; Noncheva, Z.; Petrova, M.; Dobrev, T.; Mirkova, L.; Vermeersch, R.; Demaerel, J.-P. Insoluble Anodes Used in Hydrometallurgy. *Hydrometallurgy* **2000**, *57*, 109–124. [CrossRef]
22. Barmi, M.J.; Nikoloski, A.N. Electrodeposition of Lead–Cobalt Composite Coatings Electrocatalytic for Oxygen Evolution and the Properties of Composite Coated Anodes for Copper Electrowinning. *Hydrometallurgy* **2012**, *129–130*, 59–66. [CrossRef]
23. Tobosque, P.; Maril, M.; Maril, Y.; Camurri, C.; Delplancke, J.L.; Delplancke, M.P.; Rodríguez, C.A.; Carrasco, C. Electrodeposition of Lead–Cobalt Anodes: The Effect of Electrolyte PH on Film Properties. *J. Electrochem. Soc.* **2017**, *164*, D621–D625. [CrossRef]
24. Isakhani-Zakaria, M.; Allahkaram, S.R.; Ramezani-Varzaneh, H.A. Evaluation of Corrosion Behaviour of Pb-Co_3O_4 Electrodeposited Coating Using EIS Method. *Corros. Sci.* **2019**, *157*, 472–480. [CrossRef]
25. Roy, S.; Majumder, S.B. Recent Advances in Multiferroic Thin Films and Composites. *J. Alloys Compd.* **2012**, *538*, 153–159. [CrossRef]

26. Xing, Y.T.; Micklitz, H.; Herrera, W.T.; Rappoport, T.G.; Baggio-Saitovitch, E. Superconducting Transition in Pb/Co Nanocomposites: Effect of Co Volume Fraction and External Magnetic Field. *Eur. Phys. J. B* **2010**, *76*, 353–357. [CrossRef]
27. Xing, Y.T.; Micklitz, H.; Rappoport, T.G.; Milošević, M.V.; Solórzano-Naranjo, I.G.; Baggio-Saitovitch, E. Spontaneous Vortex Phases in Superconductor-Ferromagnet Pb-Co Nanocomposite Films. *Phys. Rev. B Condens. Matter Mater. Phys.* **2008**, *78*, 224524. [CrossRef]
28. *SAINT*, version 2017.3-0; Part of Bruker APEX3 Software Package; Bruker AXS: Billerica, MA, USA, 2017.
29. *SADABS*, version 2017.3-0; Part of Bruker APEX3 Software Package; Bruker AXS: Billerica, MA, USA, 2017.
30. Sheldrick, G.M. SHELXT—Integrated Space-Group and Crystal-Structure Determination. *Acta Crystallogr. A Found. Adv.* **2015**, *71*, 3–8. [CrossRef]
31. Sheldrick, G.M. Crystal Structure Refinement with SHELXL. *Acta Crystallogr. C Struct. Chem.* **2015**, *71*, 3–8. [CrossRef]
32. Dolomanov, O.V.; Bourhis, L.J.; Gildea, R.J.; Howard, J.A.K.; Puschmann, H. OLEX2: A Complete Structure Solution, Refinement and Analysis Program. *J. Appl. Crystallogr.* **2009**, *42*, 339–341. [CrossRef]

Disclaimer/Publisher's Note: The statements, opinions and data contained in all publications are solely those of the individual author(s) and contributor(s) and not of MDPI and/or the editor(s). MDPI and/or the editor(s) disclaim responsibility for any injury to people or property resulting from any ideas, methods, instructions or products referred to in the content.

Article

Transition Metals Meet Scorpiand-like Ligands

Salvador Blasco *, Begoña Verdejo, María Paz Clares * and Enrique García-España

Instituto de Ciencia Molecular, Departamento de Química Inorgánica, Universidad de Valencia, C/Catedrático José Beltrán Martínez, 2, 46980 Paterna, Valencia, Spain; begona.verdejo@uv.es (B.V.); enrique.garcia-es@uv.es (E.G.-E.)
* Correspondence: salvador.blasco@uv.es (S.B.); m.paz.clares@uv.es (M.P.C.)

Abstract: Scorpiand-like ligands combine the preorganization of the donor atoms of macrocycles and the degrees of freedom of the linear ligands. We prepared the complexes of several of these ligands with transition metal ions and made a crystallographic and water solution speciation studies. The analysis of the resulting crystal structures show that the ligands have the ability to accommodate several metal ions and that the coordination geometry is mostly determined by the ligand. Ligand 6-[3,7-diazaheptyl]-3,6,9–triaza-1-(2,6)-pyridinacyclodecaphane (**L3**) is an hexadentate ligand that affords a family of isostructural crystals with Cu(II), Mn(II), Ni(II) and Zn(II). The attempts to obtain Co(II) crystals afforded the Co(III) structures instead. Ligand 6-[4-(2-pyridyl)-3-azabutyl]-3,6,9-triaza-1-(2,6)-pyridinacyclodecaphane (**L2**) is very similar to **L3** and yields structures similar to it, but its behavior in solution is very different due to the different interaction with protons. Ligand 6-(2-aminoethyl)-3,6,9–triaza-1-(2,6)-pyridinacyclodecaphane (**L1**) is pentadentate and its complexes allow the metal to be more accessible from the solvent. A Zn(II) structure with **L1** shows the species [ZnBrH**L1**]$^{2+}$, which exists in a narrow pH range.

Keywords: transition metals; metal ions; coordination chemistry; isostructural; scorpiand-like ligands

---

Citation: Blasco, S.; Verdejo, B.; Clares, M.P.; García-España, E. Transition Metals Meet Scorpiand-like Ligands. *Crystals* **2023**, *13*, 1338. https://doi.org/10.3390/cryst13091338

Academic Editor: Jesús Sanmartín-Matalobos

Received: 28 July 2023
Revised: 19 August 2023
Accepted: 29 August 2023
Published: 1 September 2023

Copyright: © 2023 by the authors. Licensee MDPI, Basel, Switzerland. This article is an open access article distributed under the terms and conditions of the Creative Commons Attribution (CC BY) license (https://creativecommons.org/licenses/by/4.0/).

1. Introduction

Macrocyclic ligands offer an interesting framework for the study of coordination chemistry because all their donor atoms come preorganized, arranged geometrically in space. When they coordinate transition metals, the resulting coordination geometry is determined by the combined action of the preferred coordination geometry of the metal and the ability of the ligand to adapt to it. In the case of macrocyclic ligands, they tend to impose their coordination mode. If the ligand is open, linear and flexible, it has many more conformational degrees of freedom and, therefore, it is able to adapt to a wider variety of coordination preferences. It is possible to combine both types in a family of ligands that features a macrocyclic core decorated with a flexible pending arm containing additional donor atoms. Such ligands are often referred to as scorpiands due to its similarity with the body of a scorpion, where the macrocycle would be the scorpion body and the pending arm would be the tail with a sting at the end of it. They were first reported by Lotz and Kaden [1] although the name scorpiand was first coined by Fabrizzi and collaborators [2].

In our research group, we work with a family of ligands of this kind. Some of them are derivatives of **L1** (see Figure 1), which can be readily obtained from a modified Richman–Atkins condensation of 2,6-bis(bromomethyl)pyridine with tris[2-(*N*-tolylsulphonylaminoethyl)]-amine followed by detosylation with HBr/acetic acid to obtain the hydrobromide salt [3]. **L1** consists of a 12-membered ring with four nitrogen donor atoms: one is a pyridine nitrogen, two of them are secondary amino groups, and the last one is a tertiary amino group. This macrocycle is too small to equatorially coordinate a metal atom. Instead, when a metal atom goes into the cavity, the macrocycle folds leave two adjacent coordination vacancies as it can be seen in the literature in many examples that contain either **L1** or the analog macrocycle without the pending arm [4–7]. This tendency to fold makes the bonds around N2 and N4 (see numbering in Figure 1) stretched out if it is possible. This feature is relevant in order to understand the structures presented in this work. From the synthetic point of view, **L1** contains a primary amino group, which allows to easily introduce additional moieties in that position. A sizable number of derivatives of **L1** containing different groups in that position have been synthesized and reported by our group [3,8–10]. When **L1** is reacted with 2-pyridinecarboxaldehyde, we obtain the ligand **L2**, which incorporates a sixth donor nitrogen atom in the pending arm. If the reagent is 3-bromopropylphthalimide followed by phthalimide cleavage with hydrazine, we obtain **L3**, which also features a sixth donor atom but is much more flexible than the analog **L2**.

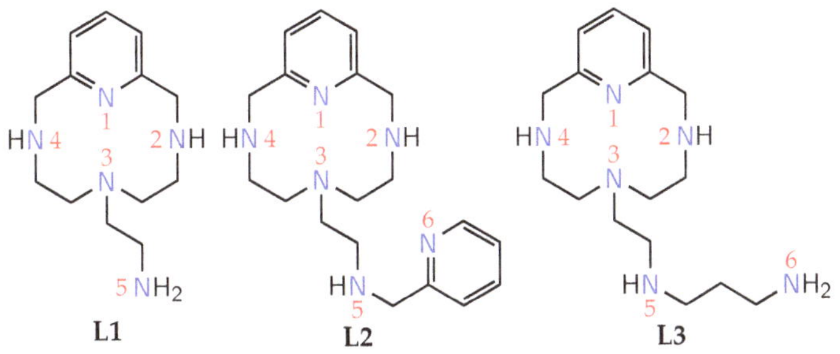

Figure 1. Ligands used in this work and nitrogen atom numbering.

These ligands are able to coordinate many types of metal atoms, mainly transition and post-transition metal ions. So far, we reported interactions with Cu, Mn, Zn, Fe and Pb [3,9,11–13], and research is underway with other metal ions as well. We focused on the copper and manganese complexes of these ligands because they show very interesting redox properties. When the metal in question is able to catalyze one-electron redox reactions, such as Cu(II) or Mn(II), we report that such complexes are able to mimic the redox cycle of superoxide dismutase (SOD), catalyzing the dismutation of the superoxide anion [8,9,11] Iron complexes of **L2** undergo oxidative dehydrogenation by the effect of a pH change [13]. These kinds of metal complexes might seem simple at first glance, but they exhibit a rather complex chemistry. Herein, we aim to present some of the features of this family of complexes by showing both solution and solid-state data in order to illustrate the interesting coordination chemistry that they exhibit.

2. Materials and Methods

2.1. Synthesis and Materials Used

The synthesis of **L1**, **L2** and **L3** was carried out as described in the literature [3,8,9] For crystal preparation and potentiometric measurements, stock solutions ca. 0.1 M of either $MnSO_4 \cdot H_2O$, $Cu(ClO_4)_2 \cdot 6H_2O$, $Ni(ClO_4)_2 \cdot 6H_2O$, $Zn(ClO_4)_2 \cdot 6H_2O$ or $CoCl_2 \cdot 2H_2O$ were prepared in Mili-Q water. The metal concentration was standardized by a suitable complexation volumetry method. In the particular case of iron stock solutions, they were prepared from $FeSO_4 \cdot 7H_2O$ in 0.1 M HCl. In these cases, both the metal concentration and the acid concentration were determined by volumetric titration. The complexes were prepared by the simple mixing of equimolar amounts of both the metal stock solution and the desired ligand in an aqueous solution. Single crystals were prepared by the slow evaporation of a pH neutral water solution of the corresponding complex. **Caution:** *Perchlorates are potentially explosive if they come in contact with organic matter, and they should be handled with extreme care.*

2.2. X-ray Single Crystal Diffractometry

Single crystal X-ray diffraction was carried out with either of the following machines: (1) Enraf-Nonius KAPPA CCD single-crystal diffractometer with MoK$_\alpha$ radiation ($\lambda = 0.71073$ Å) and a graphite monochromator with data collection by COLLECT software with utilities DENZO and SCALEPACK, or (2) Agilent Super-Nova Diffractometer with

MoK$_\alpha$ radiation ($\lambda = 0.71073$ Å) with a mirror monochromator located at the Institute of Molecular Sciences (ICMOL) from the University of Valencia, and data acquisition and treatment performed with CRYSALISPRO software [14]. If an absorption correction needed to be applied, either a spherical absorption correction as implemented in DENZO, or a semi-empirical absorption correction (MULABS [15] as implemented in PLATON [16]) was applied.

Crystals were measured at room temperature (293 K). The structures were solved initially with SHELXS [17] with direct methods. Then, they were refined with SHELXL [17] using OLEX2 [18] as the frontend. Final figures for publishing were produced using CHIMERA [19] or MERCURY [20].

2.3. Potentiometric Measurements

The potentiometric titrations were carried out in a thermostated bath at 298.1 K, using NaClO$_4$ 0.15 M as the supporting electrolyte, in the pH range 2.5–11.0. The experimental procedure (burette, potentiometer, cell, stirrer, microcomputer, etc.) is fully described elsewhere [21]. The acquisition of the electromotive force *emf* data was performed with the computer program PASAT [22,23]. The reference electrode was an Ag/AgCl electrode in saturated KCl solution. The glass electrode was calibrated as a hydrogen ion concentration probe by titration of previously standardized amounts of HCl with carbonate-free NaOH solutions and the equivalent point determined by the Gran's method [24,25], which gives the standard potential E° and the ionic product of water (pK_w = 13.73(1)). At least two reproducible titrations for each system were recorded. The computer program HYPERQUAD [26] was used to fit protonation and stability constants. The independent titration curves were treated first as separate curves, and then the curves for each system (at least two) were then merged and treated as a single set without significant variations in the values of those treated first as separate curves and then merged and treated as a single set without stability constants. The final values were those obtained from the simultaneous treatment of the merged curves. The HYSS [27] program was used to generate the distribution diagrams.

3. Results and Discussion

3.1. Interaction of L3 with Mn, Fe, Co, Cu, Ni and Zn

L3 has six donor atoms that can be geometrically arranged in an octahedral disposition. When a metal atom is placed inside the ligand cavity, the ligand itself will impose its preferred coordination geometry rather than the metal. We prepared crystals of **L3** with several transition and post-transition metals and successfully solved the structures for Mn(II), Cu(II), Ni(II) and Zn(II) with ClO$_4^-$ counterions. We named these structures **C1** to **C4**, respectively. Unfortunately, crystals of **L3** with iron could not be obtained. Also, for the case of the Co(II) complex, the Co(III) structure was obtained instead. This ligand readily coordinates metal ions. We studied some metal atoms from the first transition series for their biological relevance. We previously reported copper- or manganese-related crystal structures [9–11], but we include some of them here again for comparative purposes.

We studied metal complex formation in solution and we fitted the values of the formation constants. The results are summarized in Table 1. The stability constant values follow the expected order: Mn < Fe < Cu > Zn. All of them form stable complexes with **L3**. In addition to that, several protonation constants for each complex were also found and refined with up to three protons. The values of the first protonation constants lie between 4.59 for Cu(II) and 8.14 for Fe(II). This indicates that above pH ca. 8, the complex should be completely hexacoordinated. Distribution diagrams show that the species [ML]$^{2+}$ is completely formed in those conditions. Indeed, we obtained all the crystals for **L3** from moderately alkaline solutions. Also, some hydroxylated species were also found in the speciation study, but they only predominate at pH > 11. Solution studies with Ni(II) were also carried out, but the slow kinetics prevented the fitting of reliable formation constants. Nonetheless, suitable single-crystals were obtained from the solutions used in potentiometric measurements. Similarly, solution studies with Co(II) were also carried out, but the values obtained do not match the expected results probably due to the oxidation that happens during the titration, preventing the collection of reliable data.

Table 1. Formation constants for **L3** with some metal ions from the first transition series.

Reaction [1]	Mn(II) [2]	Fe(II)	Cu(II)	Zn(II)
$MH_2L + H \rightleftharpoons MH_3L$	—	6.35(7) [3]	3.1(2)	4.26(6)
$MHL + H \rightleftharpoons MH_2L$	5.79(9)	5.87(5)	3.59(7)	5.32(3)
$ML + H \rightleftharpoons MHL$	7.74(1)	8.17(2)	4.47(2)	6.27(3)
$M + L \rightleftharpoons ML$	10.93(1)	13.29(4)	22.90(2)	17.72(3)
$ML + H_2O \rightleftharpoons LMOH + H$	−10.74(2)	—	−10.64(4)	−10.53(4)

[1] Charges omitted. [2] Values from ref [9]. [3] Values represent $\log_{10} K$, values in parentheses represent standard deviation of the last significant figure.

Crystals **C1–C4** are isostructural (Figure 2a–d) with the same orthorhombic crystal system, P*bca* space group and atom arrangement, with the only difference being the metal atom coordinated to the ligand. Bond distances and angles are shown in Table A4. We always find that the longest bond distances involve N2 and N4. These two donor atoms are the ones that have to stretch more in order to coordinate a metal located inside the macrocycle. When the metal in question displays Jahn–Teller distortion, such as Cu(II) (crystal **C2**, Figure 2b), then the elongated axis is N2—M—N4 so that the bonds around N2 and N4 are in a more relaxed conformation. Some others, like Zn(II), do not have Jahn–Teller distortion but still have the axis N2—M—N4 as being slightly longer. This indicates that the coordination geometry is to a large extent imposed by the ligand and only to a lesser extent by the metal atom. The metal atom, though, also exerts some influence; that is why the length difference between the axial and equatorial bonds is larger for Cu(II) than for Zn(II). In these four structures, the packing is mainly ionic. In Figure 2e, details of the packing of **C1** show the arrangement of anions and cations in the unit cell. It is worthwhile noting that in all these structures, the metal ion is completely wrapped and enclosed by the ligand, leaving no vacant coordination sites and restricted access to the metal from the solvent. This is best seen in Figure 2f, where the spacefill representation shows that the metal ion is not accessible from the solvent. This feature explains that Mn(II) complexes are not oxidized in alkaline conditions open to air, and crystals containing Mn(II) are obtained. When crystals are prepared from a neutral solution of $[MnL1]^{2+}$, then a structure containing Mn(III) is obtained, which consists of two units of $[MnL1]^{3+}$ bridged by an oxo anion [28].

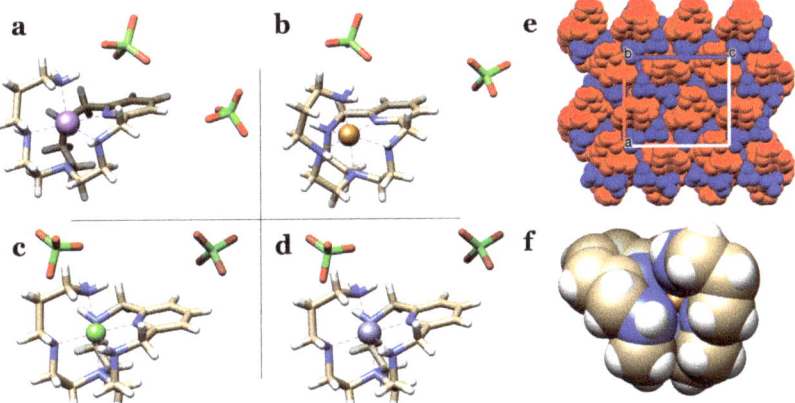

Figure 2. (**a–d**) Asymmetric unit of structures **C1–C4** featuring Mn(II), Cu(II), Ni(II) or Zn(II), respectively. (**e**) Packing of **C1** with $[CuL3]^{2+}$ depicted in red, and perchlorate anions in blue. (**f**) Spacefill of **C4** highlighting the inaccessibility of the metal atom from the solvent.

3.2. Cobalt Complexes of **L1** and **L3**

When a pale pink, diluted solution of [CoL3]$^{2+}$ in aqueous NaClO$_4$ was allowed to slowly evaporate, deep blue crystals of [CoL3](ClO$_4$)$_3$(H$_2$O) were obtained instead of the expected Co(II) ones. As we noted above, the metal atom inside the cavity was completely enclosed by the ligand, which provides protection against air oxidation for the Mn(II) complexes. This seems not to be the case for Co(II), as we consistently obtained the crystal with Co(III). As mentioned above, speciation studies under argon were carried out, but the obtained formation constants are not reported herein, as the oxidation appears to happen to some extent during the titration process, which makes the values unreliable. A similar thing also happens when Co(II) complexes of **L1** are allowed to slowly evaporate: single crystals containing the [CoBrL1]$^{2+}$ complex are obtained. Two different structures containing this complex were resolved: firstly, crystal **C6** with formula [CoBrL1]Br$_2$(H$_2$O), and crystal **C7** with formula [CoBrL1]Na(ClO$_4$)$_3$ in which sodium perchlorate is co-crystallized along with the complex. In all three examples, the Co(II) atom is easily oxidized to Co(III) by exposure to the atmosphere. The geometry of the complex [CoBrL1]$^{2+}$ from **C6** and that of the complex [CoL3]$^{3+}$ from **C5** are shown in Figure 3a,b. The complex from [CoBrL1]$^{2+}$ from **C7** is not shown, but it is identical to that from **C6**. In all three structures, the Co(III) complex displays an octahedral coordination geometry, fairly regular with angles close to 90° (see Table A5). In the case of **C6** and **C7**, the sixth coordination site is occupied by a bromide atom, which exhibits a longer bond than the others (ca. 2.3 Å), which is to be expected since it is an atom bigger than the other ones. The complex from structure **C7** is essentially identical to that of **C6** with bonds and angles that are only slightly different (see Table A5). As occurs for **C1–C4**, the metal atom is deeply embedded, and it has no access from the solvent. In this case, a single water molecule is found in the asymmetric unit along with three perchlorate anions. Structure **C6** contains a rich network of hydrogen bonds involving the hydrogen atoms of the amino groups, the water moiety and the bromide atoms (see Figure 3c). This structure is not purely electrostatic, but also the hydrogen bonds contribute to the packing. The water molecule is placed between two bromide atoms, Br2 and Br3, to which it donates hydrogen bonds, while not accepting from any other moiety. All N—H groups are hydrogen bond donors towards the nearest bromide anion. **C7** is similar to **C6**, but it features an anionic moiety of (Na(ClO$_4$)$_3$)$_n^{2n-}$ that grows along the *b* crystallographic axis. Units of [CoBrL1]$^{2+}$ cations are placed in alternating orientation, filling in the gaps (see Figure 3d). This structure has a bigger electrostatic contribution to the packing but still some relevant hydrogen bonds are found (see Table A8).

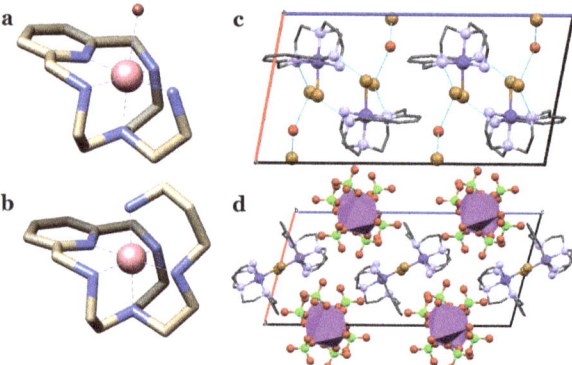

Figure 3. (**a**) Details of the complex [CoBrL1]$^{2+}$ from **C6**. Hydrogen atoms not shown. (**b**) Detail of the complex [CoL3]$^{3+}$ from **C5**. Hydrogen atoms not shown. (**c**) Unit cell of **C6** shown along the *b*-axis, displaying the packing and the internal hydrogen bonding structure. Hydrogen atoms not shown. (**d**) Unit cell of **C7** shown along the *b*-axis with Na atoms displayed in polyhedral, carbon atoms in capped stick representation, and other atoms displayed in ball-and-stick. Hydrogen atoms not shown.

*3.3. Complexes of **L2** with Cu(II) and Mn(II)*

From the coordination point of view, **L2** and **L3** are similar because the two of them are hexadentate with the donor atoms in the same positions with the same number of bonds separating one from another. The only difference is that the donor N of **L2** is on a rigid aromatic group and, for **L3**, it is located on a flexible aliphatic amino group. Indeed, the slow evaporation of an aqueous solution of the ligand with one equivalent of either Cu(ClO$_4$)$_2$ or Mn(ClO$_4$)$_2$ yields single crystals of either [ML2](ClO$_4$)$_2$ (**C1** and **C2**) and [ML3](ClO$_4$)$_2$ (**C8** and **C9**). The coordination mode is almost identical, as it can be seen comparing Figure 2a,b with Figure 4a,b. **C1**, **C2**, **C8** and **C9** crystallize in an orthorhombic system with very similar asymmetric units containing only one molecule of the complex, two perchlorate anions and no water molecules. The space group is different: **C1** and **C2**, which contain **L3**, have a P*bca* space group, while structures **C8** and **C9** have P*na*2$_1$. Indeed, **C8** is isostructural with **C9** as **C1** is to **C2**. This indicates that the metal exerts little influence outside the complex with regard to the packing. Another similarity is that the packing is mainly electrostatic as it can be seen in Figure 4c. The bond distances and angles are listed in Table A6. We observe, as we did for **L3**, the recurring feature that the longest metal–nitrogen bond distance involves N2 and N4. This is very clear for **C8**, which contains Cu(II). For **C9**, which contains Mn(II), the bond Mn1—N5 is slightly longer than Mn1—N4. A possible explanation for this is that the angles for **C9** are more distorted and further from 90° than for **C8**.

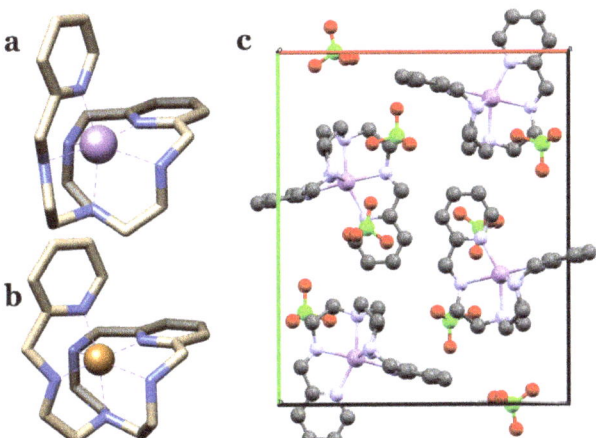

Figure 4. (**a**) Details of the complex [MnL2]$^{2+}$ from **C8**. (**b**) Details of the complex [CuL2]$^{2+}$ from **C9**. (**c**) Unit cell of **C8** viewed down *c* axis. Hydrogen atoms omitted. Mn (purple), Cl (green), O (red), N (blue), C (gray).

If we explore the chemistry in solution, the similarities end, and we see very different behavior compared to the crystal state. Above, in Table 1, we show the formation constants of **L3** with Cu(II). If we study the speciation of **L2** with Cu(II), a single constant is found, with value $\log_{10} K_{CuL} = 22.6(1)$ [29]. This value is very close to that from **L3** ($\log_{10} K_{CuL} = 22.93(2)$, see Table 1). Only this single constant is found, and its large value makes it possible for the species [CuL]$^{2+}$ to be formed at low pH values and predominate in a very large pH range (Figure 5left). This is also the reason for the larger error in the constant: only a small set of points at low pH values contain meaningful information about the formation of the complex (see Figure 5left). Nonetheless, the obtained value is confirmed by means of titration with a competing ligand in order to increase the number of data points [29]. For **L3**, on the other hand, even though the formation constant with Cu(II) is close to that of **L2**, the primary amino group can protonate more easily. As a result, the Cu(II) atom is introduced at ca. one unit of pH higher, and the species [CuL3]$^{2+}$ is not completely formed until pH ~ 6 (Figure 5right).

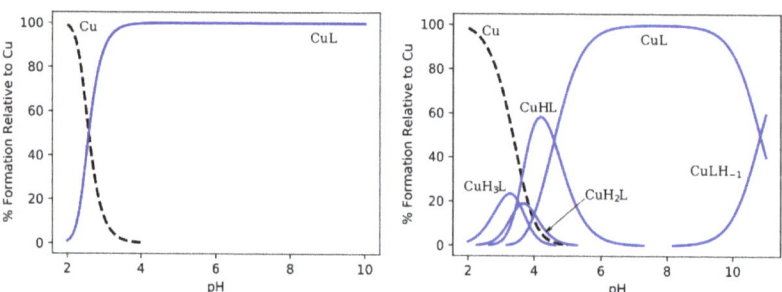

Figure 5. Species distribution diagram for the system Cu(II) with **L2** (**left**) and **L3** (**right**) with [Cu^{2+}] = [L] = 1 × 10^{-3} M. Charges omitted.

3.4. Complex of **L1** with Zn(II)

L1 is a good ligand for many transition metal atoms, such as Cu(II) [3], Mn(II) [28] or Zn(II) [12]. Usually in all the crystal structures of **L1** reported by us, the metal atom is pentacoordinated with all donor atoms bonding the metal. Solution studies indicate that the complex [ZnL1]$^{2+}$ can take one or two protons to form species [ZnHL1]$^{3+}$ and [ZnH$_2$L1]$^{4+}$ [12].

We were able to obtain single crystals of one such species, [ZnHL1]$^{3+}$. This structure, named **C10**, is shown in Figure 6. The terminal amino group can be protonated more easily than the other amino groups, and that coordination bond can break. In this case, the tip of the tail would be +1, while the metal next to it is +2. Therefore, the tail would fold away from the metal, giving, as a result, an open conformation, with the metal much more exposed to the solvent. **C10** presents a triclinic P$\bar{1}$ system and space group, and its asymmetric unit contains the complex [ZnBrHL1]$^{2+}$ with one water molecule and one ZnBr$_4^{2-}$ counterion. In the unit of [ZnBrHL1]$^{2+}$, the Zn(II) is pentacoordinated by the nitrogen atoms of **L1**, except N5, which is protonated, bearing one positive charge and being folded away from the metal atom, thus minimizing the electrostatic repulsion. The macrocycle is only partly folded as if it was trying to equatorially coordinate the metal ion but, being too small to do it, it stops halfway. The aforementioned electrostatic repulsion may also contribute to this conformation. N5 would be the fifth donor atom, but instead, a bromide anion fills this vacancy. The units of [ZnBrHL1]$^{2+}$ are grouped in pairs that share two hydrogen bonds between N4 and Br1 (see Figure 6b). A total of six hydrogen bonds can be located around each unit of complex. The details are shown in Table A9. Both electrostatic and hydrogen bonds are the main forces in the packing. The origin of ZnBr$_4^{2-}$ is the free metal available at pH ~ 5, where the complex is not completely formed and the bromide counterions are from the **L1** hydrobromide salt.

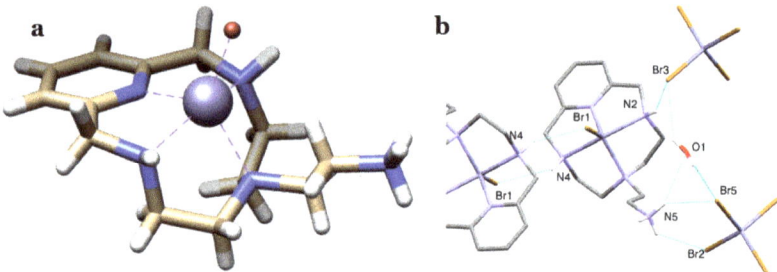

Figure 6. (**a**) Details of the complex [ZnBrHL1]$^{2+}$ from **C10**. (**b**) Hydrogen bonding network around the complex.

4. Conclusions

The ligands presented herein have the ability to coordinate a variety of metal ions. We focused on metals from the first transition series and showed examples of Cu(II), Mn(II), Co(III), and Ni(II), as well as the post-transition metal ion Zn(II). The overall packing is mostly determined by the ligand. The same ligand with different metal atoms tends to yield either isostructural or structurally similar crystals provided that the charge does not change. **L2** and **L3** have almost the same coordination geometry and, as a consequence, the structures they afford are very similar. However, solid-state structures and solution studies show different pictures about the properties of the complexes due to the different interaction of the ligands with protons as it can be seen by comparing the speciation of Cu(II) and Mn(II) with either **L2** or **L3**. **L1**, which is pentadentate, leaves the metal atom more exposed to the solvent.

Author Contributions: Conceptualization, S.B.; resources, B.V. and M.P.C.; data curation, S.B.; writing—original draft preparation, S.B.; writing—review and editing, S.B. and E.G.-E.; visualization, S.B.; supervision, E.G.-E.; funding acquisition, E.G.-E. All authors have read and agreed to the published version of the manuscript.

Funding: Financial support from the Spanish Ministerio de Economía y Competitividad (Project PID 2019-110751RD-I00) and the Conselleria de Innovación, Universidades, Ciencia y Sociedad Digital of the Generalitat Valenciana (PROMETEO Grant CIPROM/2021/030) is acknowledged.

Data Availability Statement: Crystallographic data have been deposited in the Cambridge Structural Database. These data can be obtained free of charge via http://www.ccdc.cam.ac.uk/conts/retrieving.html accessed on 25 August 2023 (or from the CCDC, 12 Union Road, Cambridge CB2 1EZ, UK; Fax: +44 1223 336033; E-mail: deposit@ccdc.cam.ac.uk). See deposition numbers in Table A1.

Conflicts of Interest: The authors declare no conflict of interest.

Abbreviations

The following abbreviations are used in this manuscript:

SOD	Superoxide dismutase
CCDC	Cambridge Crystallographic Data Center
emf	Electromotive force

Appendix A. Additional Tables

Appendix A.1. Crystallographic Data Tables

Table A1. Deposition numbers of crystal structures **C1**–**C10** in the Cambridge Crystallographic Data Center.

Structure	CCDC n.	Reference(s)
C1	1020483	[9,11]
C2	897979	[10]
C3	2281410	this work
C4	2281411	this work
C5	2281412	this work
C6	2281413	this work
C7	2281414	this work
C8	888624	[9]
C9	1423684	[29]
C10	2281415	this work

Table A2. Crystallographic data for structures **C3**–**C5**. Values in parentheses represent standard deviation of the last significant figure.

	C3	C4	C5
Formula	$C_{16}H_{30}Cl_2NiN_6O_8$	$C_{16}H_{30}Cl_2ZnN_6O_8$	$C_{16}H_{32}Cl_3CoN_6O_{13}$
d calc./g cm^{-3}	1.596	1.599	1.756
μ/mm^{-1}	1.108	1.315	1.053
Formula Weight	564.07	570.73	681.75
Colour	pale red	colourless	dark blue
Shape	prism	prism	block
Size/mm^3	0.20 × 0.11 × 0.09	0.14 × 0.11 × 0.09	0.20 × 0.10 × 0.10
Crystal System	orthorrhombic	orthorrhombic	monoclinic
Space Group	$Pbca$	$Pbca$	$P2_1/c$
a/Å	17.4035(5)	17.3510(4)	16.9810(10)
b/Å	12.6487(6)	12.7480(8)	9.2817(4)
c/Å	21.3344(9)	21.4350(9)	21.4350(9)
β/°			109.954(2)
V/Å^3	4696.4(3)	4741.2(4)	2578.5(3)
Z	8	8	4
θ_{max}/°	27.428	27.088	28.101
θ_{min}/°	3.021	3.334	1.276
Measured Refl.	9766	9615	9850
Independent Refl.	5325	5189	5759
R_{int}	0.0529	0.0580	0.0619
Parameters	298	298	356
Restraints	0	0	5
Largest Peak	0.609	0.652	1.036
Deepest Hole	−0.533	−0.528	−0.614
GooF	0.983	1.017	1.021
wR_2 (all data)	0.2460	0.2409	0.2367
wR_2	0.2004	0.1784	0.1975
R_1 (all data)	0.1464	0.1505	0.1418
R_1	0.0699	0.0684	0.0767

Table A3. Crystallographic data for structures **C6**, **C7** and **C10**. Values in parentheses represent standard deviation of the last significant figure.

	C6	C7	C10
Formula	$C_{13}H_{25}Br_3CoN_5O$	$C_{13}H_{23}BrCoN_5Cl_3NaO_{12}$	$C_{13}H_{24}BrN_5OZn_2$
d calc./g cm^{-3}	2.016	2.028	2.224
μ/mm^{-1}	7.539	2.804	10.398
Formula Weight	566.04	709.54	798.68
Colour	blue	violet	colourless
Shape	spear	block	prism
Size/mm^3	0.70 × 0.44 × 0.32	0.50 × 0.30 × 0.30	0.33 × 0.17 × 0.11
Crystal System	monoclinic	monoclinic	triclinic
Space Group	$P2_1/c$	$P2_1/c$	$P\bar{1}$
a/Å	10.2042(3)	10.7720(3)	8.1632(2)
b/Å	9.3402(2)	9.6220(6)	10.6612(3)
c/Å	19.9239(6)	23.2950(11)	14.4818(4)
α/°			107.3420(15)
β/°	100.9050(14)	105.792(3)	91.8500(16)
γ/°			96.3360(18)
V/Å^3	1864.64(9)	2323.4(2)	1192.88(6)
Z	4	4	2
θ_{max}/°	25.000	27.490	27.498
θ_{min}/°	2.416	2.285	2.017

Table A3. Cont.

	C6	C7	C10
Measured Refl.	18009	9309	26571
Independent Refl.	3282	5271	5472
R_{int}	0.1045	0.0708	0.0733
Parameters	225	333	237
Restraints	3	0	2
Largest Peak	2.032	1.615	1.372
Deepest Hole	−2.621	−1.835	−1.465
GooF	1.028	0.999	1.022
wR_2 (all data)	0.2578	0.2650	0.0983
wR_2	0.2419	0.2321	0.0881
R_1 (all data)	0.0896	0.1415	0.0671
R_1	0.0764	0.0813	0.0396

Appendix A.2. Tables of Distances and Angles

Table A4. Bond distances (Å) and angles (°) for complexes of **L3**.

Bond/Angle	C1 (M=Mn)	C2 (M=Cu)	C3 (M=Ni)	C4 (M=Zn)
M—N1	2.216(4)	2.043(6)	2.024(4)	2.110(5)
M—N2	2.298(5)	2.389(6)	2.203(5)	2.246(5)
M—N3	2.267(5)	2.103(5)	2.113(4)	2.203(5)
M—N4	2.345(5)	2.337(6)	2.179(5)	2.278(6)
M—N5	2.255(4)	2.028(6)	2.105(5)	2.150(6)
M—N6	2.213(5)	2.031(6)	2.099(5)	2.119(5)
N1—M—N2	71.80(18)	76.3(2)	79.23(19)	76.3(2)
N1—M—N3	99.98(18)	91.4(2)	94.95(17)	93.9(2)
N1—M—N4	73.15(18)	77.5(2)	79.3(2)	75.9(2)
N1—M—N5	175.81(18)	176.1(2)	178.2(2)	175.6(2)
N1—M—N6	93.77(19)	91.2(2)	90.1(2)	93.3(2)
N2—M—N3	78.23(18)	79.3(2)	81.80(18)	81.2(2)
N2—M—N4	133.93(19)	147.1(2)	152.77(19)	144.9(2)
N5—M—N2	111.05(19)	103.0(2)	100.94(19)	103.2(2)
N6—M—N2	106.19(19)	97.2(2)	96.2(2)	103.5(2)
N3—M—N4	79.42(18)	81.9(2)	83.53(18)	79.7(2)
N5—M—N3	77.9(2)	84.7(2)	83.28(19)	81.7(2)
N6—M—N3	166.25(19)	175.0(3)	174.1(2)	172.2(2)
N5—M—N4	102.83(17)	101.9(2)	100.0(2)	102.9(2)
N6—M—N4	105.03(19)	102.8(2)	100.43(19)	99.1(2)
N6—M—N5	88.38(19)	92.7(2)	91.7(2)	91.1(2)

Table A5. Bond distances (Å) and angles (°) for complexes of the Co(III) structures. Values in parentheses represent standard deviation of the last significant figure.

C5		C6		C7	
Co1—N1	1.913(4)	Co1—N1	1.889(8)	Co1—N1	1.892(6)
Co1—N2	2.017(4)	Co1—N2	1.983(8)	Co1—N2	2.007(6)
Co1—N3	1.962(4)	Co1—N4	1.967(8)	Co1—N3	1.985(6)
Co1—N4	2.026(4)	Co1—N3	1.985(8)	Co1—N4	2.011(6)
Co1—N5	2.012(4)	Co1—N5	1.951(9)	Co1—N5	1.956(6)
Co1—N6	1.950(5)	Co1—Br4	2.3371(19)	Co1—Br1	2.3600(13)

Table A5. *Cont.*

C5		C6		C7	
N1—Co1—N4	82.03(17)	N1—Co1—Br4	89.1(2)	N1—Co1—Br1	87.07(17)
N1—Co1—N2	83.91(17)	N1—Co1—N2	83.1(3)	N1—Co1—N2	83.1(3)
N1—Co1—N3	96.97(18)	N1—Co1—N4	97.0(3)	N1—Co1—N3	97.1(2)
N1—Co1—N6	84.41(19)	N1—Co1—N3	83.6(3)	N1—Co1—N4	83.0(3)
N5—Co1—N4	97.75(18)	N2—Co1—Br4	95.1(2)	N2—Co1—Br1	93.98(18)
N5—Co1—N2	96.52(18)	N4—Co1—N2	86.1(3)	N3—Co1—N2	86.4(3)
N3—Co1—N4	86.20(18)	N4—Co1—N3	86.1(3)	N3—Co1—N4	86.5(3)
N3—Co1—N2	85.41(18)	N3—Co1—Br4	94.1(2)	N4—Co1—Br1	94.24(18)
N3—Co1—N5	84.18(19)	N5—Co1—Br4	87.4(2)	N5—Co1—Br1	89.91(18)
N6—Co1—N4	91.4(2)	N5—Co1—N2	96.4(4)	N5—Co1—N2	97.7(3)
N6—Co1—N2	97.32(19)	N5—Co1—N4	86.6(3)	N5—Co1—N3	85.9(3)
N6—Co1—N5	94.4(2)	N5—Co1—N3	97.5(3)	N5—Co1—N4	96.7(3)

Table A6. Bond distances (Å) and angles (°) for complexes of **L2**. Values in parentheses represent standard deviation of the last significant figure.

C8 (M=Cu)		C9 (M=Mn)	
Cu1—N1	2.031(11)	Mn1—N1	2.165(6)
Cu1—N2	2.26(3)	Mn1—N2	2.333(16)
Cu1—N3	2.114(12)	Mn1—N3	2.256(11)
Cu1—N4	2.368(18)	Mn1—N4	2.26(2)
Cu1—N5	1.962(11)	Mn1—N5	2.266(13)
Cu1—N6	2.011(11)	Mn1—N6	2.173(10)
N1—Cu1—N2	78.7(11)	N1—Mn1—N2	73.6(6)
N1—Cu1—N3	93.2(4)	N1—Mn1—N3	102.7(3)
N1—Cu1—N4	78.6(10)	N1—Mn1—N4	72.2(7)
N2—Cu1—N4	150.8(6)	N1—Mn1—N5	166.2(6)
N3—Cu1—N2	84.5(10)	N1—Mn1—N6	103.9(3)
N3—Cu1—N4	78.6(9)	N3—Mn1—N2	81.2(6)
N5—Cu1—N1	178.7(8)	N3—Mn1—N4	76.9(7)
N5—Cu1—N2	100.5(13)	N3—Mn1—N5	79.2(5)
N5—Cu1—N3	85.7(5)	N4—Mn1—N2	133.6(4)
N5—Cu1—N4	101.8(12)	N4—Mn1—N5	95.2(7)
N5—Cu1—N6	82.4(5)	N5—Mn1—N2	120.1(7)
N6—Cu1—N1	98.8(4)	N6—Mn1—N2	111.0(7)
N6—Cu1—N2	97.2(12)	N6—Mn1—N3	153.0(4)
N6—Cu1—N3	168.0(4)	N6—Mn1—N4	106.9(7)
N6—Cu1—N4	104.3(10)	N6—Mn1—N5	73.9(5)

Appendix A.3. Tables of Hydrogen Bonds

Table A7. Hydrogen bond data for **C6**. Symmetry operations: (i) $1-x$, $-y$, $1-z$; (ii) x, $\frac{1}{2}-y$, $\frac{1}{2}+z$. Distances in Å, angles in °. Values in parentheses represent standard deviation of the last significant figure.

		Distances		Angle
D—H···A [1]	D—H	H···A	D···A	D—H···A
N5—H5A···Br2[i]	0.89	2.88	3.659(8)	146
O1—H1A···Br2[i]	0.83(13)	2.86(13)	3.355(11)	120(10)
O1—H1B···Br3[i]	0.81(6)	2.58(7)	3.368(11)	164(7)
N2—H2···Br2[i]	0.98	2.32	3.295(8)	174
N5—H5B···Br2[ii]	0.89	2.57	3.427(8)	160
N4—H4···Br2[ii]	0.98	2.44	3.384(8)	161
N5—H5C···Br2[i]	0.89	2.82	3.659(8)	157
N5—H5D···Br2[i]	0.89	2.64	3.427(8)	147

[1] D = donor, H = hydrogen, A = acceptor.

Table A8. Hydrogen bond data for **C7**. Symmetry operations: (i) $1 - x, 1 - y, 1 - z$. Distances in Å, angles in °. Values in parentheses represent standard deviation of the last significant figure.

D—H···A [1]	D—H	Distances H···A	D···A	Angle D—H···A
N2—H2A···O13[i]	0.98	2.23	3.125(10)	152
N4—H4A···Br1[i]	0.98	2.39	3.346(7)	165
N5—H5A···O13[i]	0.89	2.11	2.945(10)	155
N5—H5B···Br1[i]	0.89	2.70	3.476(7)	147

[1] D = donor, H = hydrogen, A = acceptor.

Table A9. Hydrogen bond data for **C10**. Symmetry operations: (i) $1 - x, y, z$; (ii) $1 - x, 1 - y, -z$; (iii) $1 - x, 2 - y, 1 - z$; (iv) $1 - x, 1 + y, z$. Distances in Å, angles in °. Values in parentheses represent standard deviation of the last significant figure.

D—H···A [1]	D—H	Distances H···A	D···A	Angle D—H···A
O1—H1C···Br3[i]	0.81(4)	2.62(8)	3.290(14)	141(9)
O1—H1D···Br5[ii]	0.83(11)	2.50(10)	3.324(13)	175(9)
N2—H2···Br3[i]	0.98	2.53	3.440(4)	155
N4—H4···Br1[iii]	0.98	2.71	3.550(4)	144
N5—H5A···Br2[ii]	0.89	2.67	3.364(6)	135
N5—H5B···Br4[iv]	0.89	2.48	3.316(6)	158
N5—H5C···O1	0.89	2.50	3.277(16)	147
N5—H5C···Br5[ii]	0.89	2.93	3.637(6)	138

[1] D = donor, H = hydrogen, A = acceptor.

References

1. Lotz, T.J.; Kaden, T.A. pH-Induced co-ordination geometry change in a macrocyclic nickel (II) complex. *J. Chem. Soc. Chem. Commun.* **1977**, *1*, 15–16. [CrossRef]
2. Pallavicini, P.S.; Perotti, A.; Poggi, A.; Seghi, B.; Fabbrizzi, L. N-(aminoethyl) cyclam: A tetraaza macrocycle with a coordinating tail (scorpiand). Acidity controlled coordination of the side chain to nickel (II) and nickel (III) cations. *J. Am. Chem. Soc.* **1987**, *109*, 5139–5144. [CrossRef]
3. Verdejo, B.; Ferrer, A.; Blasco, S.; Castillo, C.E.; González, J.; Latorre, J.; Máñez, M.A.; Basallote, M.G.; Soriano, C.; García-España, E. Hydrogen and copper ion-induced molecular reorganizations in scorpionand-like ligands. A potentiometric, mechanistic, and solid-state study. *Inorg. Chem.* **2007**, *46*, 5707–5719. [CrossRef] [PubMed]
4. Johnston, H.M.; Palacios, P.M.; Pierce, B.S.; Green, K.N. Spectroscopic and solid-state evaluations of tetra-aza macrocyclic cobalt complexes with parallels to the classic cobalt(II) chloride equilibrium. *J. Coord. Chem.* **2016**, *69*, 1979–1989. [CrossRef]
5. Félix, V.; Costa, J.; Delgado, R.; Drew, M.G.B.; Duarte, M.T.; Resende, C. X-Ray diffraction and molecular mechanics studies of 12-, 13-, and 14-membered tetraaza macrocycles containing pyridine: Effect of the macrocyclic cavity size on the selectivity of the metal ion. *J. Chem. Soc. Dalton Trans.* **2001**, 1462–1471. [CrossRef]
6. Magallón, C.; Griego, L.; Hu, C.H.; Company, A.; Ribas, X.; Mirica, L.M. Organometallic Ni(ii), Ni(iii), and Ni(iv) complexes relevant to carbon–carbon and carbon–oxygen bond formation reactions. *Inorg. Chem. Front.* **2022**, *9*, 1016–1022. [CrossRef]
7. Cavalleri, M.; Panza, N.; di Biase, A.; Tseberlidis, G.; Rizzato, S.; Abbiati, G.; Caselli, A. [Zinc(II)(Pyridine-Containing Ligand)] Complexes as Single-Component Efficient Catalyst for Chemical Fixation of CO_2 with Epoxides. *Eur. J. Org. Chem.* **2021**, *2021*, 2764–2771. [CrossRef]
8. Castillo, C.E.; Máñez, M.A.; Basallote, M.G.; Clares, M.P.; Blasco, S.; García-España, E. Copper (II) complexes of quinoline polyazamacrocyclic scorpiand-type ligands: X-ray, equilibrium and kinetic studies. *Dalton Trans.* **2012**, *41*, 5617–5624. [CrossRef]
9. Clares, M.P.; Serena, C.; Blasco, S.; Nebot, A.; del Castillo, L.; Soriano, C.; Domenech, A.; Sánchez-Sánchez, A.V.; Soler-Calero, L.; Mullor, J.L.; et al. Mn (II) complexes of scorpiand-like ligands. A model for the MnSOD active centre with high in vitro and in vivo activity. *J. Inorg. Biochem.* **2015**, *143*, 1–8. [CrossRef]
10. Inclán, M.; Albelda, M.T.; Frías, J.C.; Blasco, S.; Verdejo, B.; Serena, C.; Salat-Canela, C.; Díaz, M.L.; García-España, A.; García-España, E. Modulation of DNA Binding by Reversible Metal-Controlled Molecular Reorganizations of Scorpiand-like Ligands. *J. Am. Chem. Soc.* **2012**, *134*, 9644–9656. [CrossRef]
11. Clares, M.P.; Blasco, S.; Inclán, M.; del Castillo Agudo, L.; Verdejo, B.; Soriano, C.; Doménech, A.; Latorre, J.; García-España, E. Manganese (II) complexes of scorpiand-like azamacrocycles as MnSOD mimics. *Chem. Commun.* **2011**, *47*, 5988–5990. [CrossRef] [PubMed]

12. González, J.; Llinares, J.M.; Belda, R.; Pitarch, J.; Soriano, C.; Tejero, R.; Verdejo, B.; García-España, E. Tritopic phenanthroline and pyridine tail-tied aza-scorpiands. *Org. Biomol. Chem.* **2010**, *8*, 2367–2376. [CrossRef] [PubMed]
13. Clares, M.P.; Acosta-Rueda, L.; Castillo, C.E.; Blasco, S.; Jimenez, H.R.; Garcia-Espana, E.; Basallote, M.G. Iron (II) Complexes with Scorpiand-Like Macrocyclic Polyamines: Kinetico-Mechanistic Aspects of Complex Formation and Oxidative Dehydrogenation of Coordinated Amines. *Inorg. Chem.* **2017**, *56*, 4400–4412. [CrossRef]
14. *CrysAlisPro*, Version 1.171.36.28; Agilent Technologies: Santa Clara, CA, USA, 2012.
15. Blessing, R.H. An empirical correction for absorption anisotropy. *Acta Crystallogr. A* **1995**, *51*, 33–38. [CrossRef] [PubMed]
16. Spek, A.L. Structure validation in chemical crystallography. *Acta Crystallogr. D* **2009**, *65*, 148–155. [CrossRef] [PubMed]
17. Sheldrick, G.M. Crystal structure refinement with SHELXL. *Acta Crystallogr. C* **2015**, *C71*, 3–8. [CrossRef]
18. Dolomanov, O.; Bourhis, L.; Gildea, R.; Howard, J.; Puschmann, H. Olex2—A complete package for Molecular Crystallography. *J. Appl. Crystallogr.* **2009**, *42*, 339–342. [CrossRef]
19. Pettersen, E.F.; Goddard, T.D.; Huang, C.C.; Couch, G.S.; Greenblatt, D.M.; Meng, E.C.; Ferrin, T.E. UCSF Chimera—A visualization system for exploratory research and analysis. *J. Comput. Chem.* **2004**, *25*, 1605–1612. [CrossRef]
20. Macrae, C.F.; Sovago, I.; Cottrell, S.J.; Galek, P.T.; McCabe, P.; Pidcock, E.; Platings, M.; Shields, G.P.; Stevens, J.S.; Towler, M.; et al. Mercury 4.0: From visualization to analysis, design and prediction. *J. Appl. Crystallogr.* **2020**, *53*, 226–235. [CrossRef]
21. Garcia-España, E.; Ballester, M.J.; Lloret, F.; Moratal, J.M.; Faus, J.; Bianchi, A. Low-spin six-coordinate cobalt(II) complexes. A solution study of tris(violurato)cobaltate(II) ions. *J. Chem. Soc. Dalton Trans.* **1988**, 101–104. [CrossRef]
22. Fontanelli, M. In *Proceedings of the I Spanish-Italian Congress on Thermodynamics of Metal Complexes, Castellón, Spain, 1990*; Diputación de Castellón: Castellón, Spain, 1990.
23. Savastano, M.; Fiaschi, M.; Ferraro, G.; Gratteri, P.; Mariani, P.; Bianchi, A.; Bazzicalupi, C. Sensing Zn^{2+} in Aqueous Solution with a Fluorescent Scorpiand Macrocyclic Ligand Decorated with an Anthracene Bearing Tail. *Molecules* **2020**, *25*, 1355. [CrossRef] [PubMed]
24. Gran, G. Determination of the equivalence point in potentiometric titrations. Part II. *Analyst* **1952**, *77*, 661–671. [CrossRef]
25. Rossotti, F.; Rossotti, H. Potentiometric titrations using Gran plots: A textbook omission. *J. Chem. Ed.* **1965**, *42*, 375. [CrossRef]
26. Gans, P.; Sabatini, A.; Vacca, A. Investigation of equilibria in solution. Determination of equilibrium constants with the HYPERQUAD suite of programs. *Talanta* **1996**, *43*, 1739–1753. [CrossRef]
27. Alderighi, L.; Gans, P.; Ienco, A.; Peters, D.; Sabatini, A.; Vacca, A. Hyperquad simulation and speciation (HySS): A utility program for the investigation of equilibria involving soluble and partially soluble species. *Coord. Chem. Rev.* **1999**, *184*, 311–318. [CrossRef]
28. Blasco, S.; Cano, J.; Clares, M.P.; García-Granda, S.; Doménech, A.; Jiménez, H.R.; Verdejo, B.; Lloret, F.; García-España, E. A Binuclear MnIII Complex of a Scorpiand-Like Ligand Displaying a Single Unsupported MnIII–O–MnIII Bridge. *Inorg. Chem.* **2012**, *51*, 11698–11706. [CrossRef]
29. Blasco, S.; Verdejo, B.; Clares, M.P.; Castillo, C.E.; Algarra, A.G.; Latorre, J.; Máñez, M.A.; Basallote, M.G.; Soriano, C.; García-España, E. Hydrogen and Copper Ion Induced Molecular Reorganizations in Two New Scorpiand-Like Ligands Appended with Pyridine Rings. *Inorg. Chem.* **2010**, *49*, 7016–7027. [CrossRef]

Disclaimer/Publisher's Note: The statements, opinions and data contained in all publications are solely those of the individual author(s) and contributor(s) and not of MDPI and/or the editor(s). MDPI and/or the editor(s) disclaim responsibility for any injury to people or property resulting from any ideas, methods, instructions or products referred to in the content.

Article

Sulfonato Complex Formation Rather than Sulfonate Binding in the Extraction of Base Metals with 2,2′-Biimidazole: Extraction and Complexation Studies

Pulleng Moleko-Boyce *, Eric C. Hosten and Zenixole R. Tshentu *

Department of Chemistry, Nelson Mandela University, P.O. Box 77000, Gqeberha 6031, South Africa
* Correspondence: pulleng.moleko-boyce@mandela.ac.za (P.M.-B.); zenixole.tshentu@mandela.ac.za (Z.R.T.); Tel.: +27-41-504-1359 (P.M.-B.); +27-41-504-2074 (Z.R.T.)

Abstract: The application of a bidentate aromatic N,N'-donor ligand, 2,2′-biimidazole (BIIMH$_2$), as an extractant in the form of 1-octyl-2,2′-biimidazole (OBIIMH) and related derivatives in the solvent extraction of base metal ions (Mg^{2+}, Mn^{2+}, Fe^{3+}, Fe^{2+}, Co^{2+}, Ni^{2+}, Cu^{2+} and Zn^{2+}) from an acidic sulfonate medium using dinonylnaphthalene disulfonic acid (DNNDSA) as a synergist was investigated. OBIIMH with DNNDSA as a co-extractant showed a lack of selectivity for base metals ions (Mg^{2+}, Mn^{2+}, Fe^{3+}, Fe^{2+}, Co^{2+}, Ni^{2+}, Cu^{2+} and Zn^{2+}) despite its similarity with a related bidentate aromatic ligand, 2,2′-pyridylimidazole, which showed preference for Ni(II) ions. The nickel(II) specificity, through stereochemical "tailor-making", was not achieved as expected and the extracted species were isolated to study the underlying chemistry. The homemade metal sulfonate salts, M(RSO$_3$)$_2$·6H$_2$O (R = Toluene and M^{2+} = Co^{2+}, Ni^{2+}, Cu^{2+} and Zn^{2+}), were used as precursors of the metal complexes of BIIMH$_2$ using toluene-4-sulfonic acid as the representative sulfonate. Spectroscopic analysis and single-crystal X-ray analysis supported the formation of similar neutral distorted octahedral sulfonato complexes through the *bis* coordination of BIIMH$_2$ and two sulfonate ions rather than the formation of cationic complex species with anion coordination of sulfonates. We attributed the observation of similar complex species and the similar stability constants of the *bis*-complexes in solution as the cause for the lack of pH-metric separation of the later 3d metal ions.

Keywords: solvent extraction; 2,2′-biimidazole; coordination chemistry; crystal structures

1. Introduction

The application of amine extractants in the neutral form has not been extensively explored as separating agents for base metal ions from a basic bonding viewpoint [1]. Strong ligands with an O-donor-only character show a lack of relative preference for the base metal ions while nitrogenous ligands show promise. Aromatic nitrogenous ligands have a relative preference for metal ions which could relate to the possibility of σ and π bonding [2]. Imidazole ligands/extractants show high-formation constants with later 3d-transition metals [1,3], resulting in high extraction efficiencies and interactions with these metals in slightly strongly to weakly acidic media since their protonation constants are not too high or too low. However, large counterions such as organic sulfonates which act as synergists in the extraction of cationic complexes are frequently employed for a solvent extraction system to facilitate the transfer of the complexes efficiently to the organic phase [4].

A bidentate ligand, 2,2′-biimidazole (Figure 1A), has been used for the extraction of base metals in this study. The high complex formation offered by the bidentate ligand and the low protonation constant of the imidazole group compared with aliphatic amines allows for the formation of the inner sphere complexes in a highly acidic medium. These characteristics were to be exploited in this study in an analogous matter to the use of 2,2′-pyridylimidazole that we have studied previously [3]. It is anticipated that specificity for

base metal ions could be achieved through stereochemical "tailor-making", i.e., through the formation of complexes of different but preferred geometries. The outcome of this particular study is rather surprising, and we attempt to explain it from a basic chemistry point of view. Sulfonates are typically used as synergistic counterions to extract cationic complexes in solvent-extraction systems. However, this account presents their non-innocent nature towards inner sphere rather than outer sphere coordination in the complex formation of extracted species in a solvent-extraction system.

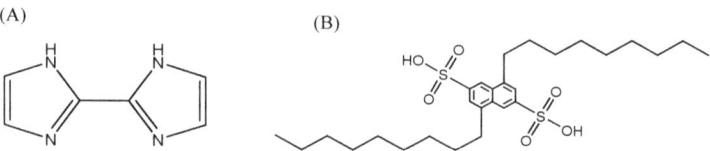

Figure 1. The chemical structures of (**A**) 2,2′-biimidazole (BIIMH$_2$) and (**B**) dinonlynapthalene disulfonic acid (DNNDSA).

The alkylated derivatives of the bidentate N,N-donor 2,2′-biimidazole ligand were investigated for their selectivity for nickel(II) from other base metals along with dinonylnaphthalene disulfonic acid (DNNDSA) as a synergist (Figure 1B) in a solvent-extraction system. The conditions for the extraction studies were designed using the OBIIMH (octyl derivative) as an extractant and DNNDSA as a synergist; both were dissolved in 80% 2-octanol and 20% Shellsol 2325 as diluent and modifier, respectively. The underlying coordination chemistry was investigated through stability constants studies and molecular structures of model-extracted species via spectroscopic techniques and single crystal X-ray crystallography. Herein, we concluded on the non-innocent nature of the sulfonates as synergistic counterions by showing evidence of inner sphere coordination.

2. Materials and Methods

2.1. Reagents and Materials

The reagents and materials used in this study, including ammonium acetate (99%), 1-Bromodecane (99%), 1-Bromoheptane (99%), 1-Bromooctane (99%), CaSO$_4$·H$_2$O (98.5%), Co(ClO$_4$)$_2$·6H$_2$O (98%), CoSO$_4$·7H$_2$O (97.5%), CdSO$_4$·H$_2$O (98%), Cu(ClO$_4$)$_2$·6H$_2$O (98%), CuSO$_4$ (anhydrous) (99%), DNNDSA (55 wt % in Iso-butanol), Fe$_2$(SO$_4$)$_3$·xH$_2$O (70%), Fe(SO$_4$)·7H$_2$O (98%), Glyoxal (40 wt % in water), MgSO$_4$·7H$_2$O (99.7%), MnSO$_4$·7H$_2$O (99.2%), Ni(ClO$_4$)$_2$·6H$_2$O (98%), NiSO$_4$·6H$_2$O (98%), Shellsol 232), Toluene-4-sulfonic acid (98%), Zn(ClO$_4$)$_2$·6H$_2$O (98%), ZnSO$_4$·7H$_2$O (99.5%), Acetone (98%), Diethylether (99%), Ethanol (99%), Ethyl acetate (98.8%), H$_2$SO$_4$ (98%), Methanol (99.9%) and 2-Octanol (98%), were purchased from Sigma-Aldrich, South Africa. All solvents were purchased from Merch and used as received. Standard solutions of the metal ions of AAS calibration were prepared from 1000 ppm stock solutions in 0.5 M nitric acid supplied by EC lab services from South Africa.

2.2. Instrumentation

The purity and identity of the extractants were determined by using ^{1}H NMR spectroscopy on a Bruker AMX 400 NMR MHz spectrometer and reported relative to tetramethylsilane (δ 0.00). The metal complexes were characterized using infrared spectroscopy on both Perkin Elmer 400 FTIR and 100 FTIR-ATR spectrometers. The metal complexes were characterized using infrared spectroscopy and recorded on either a Perkin Elmer 400FTIR spectrometer in the mid-IR range (400–4000 cm^{-1}) as KBr pellets or as neat compounds with a Perkin Elmer 100 FTIR-ATR (650–4000 cm^{-1}) spectrometer. The solid reflectance spectra of ligands and complexes were recorded on a Shimadzu UV-VIS-NIR Spectrophotometer UV-3100 with an MPCF-3100 sample compartment with samples mounted between two quartz discs that fit into a sample holder coated with barium sulfate. The spectra were recorded over the wavelength range of 250–1400 nm, and the scans were conducted at a medium speed using a 20 nm slit width.

Elemental analysis was carried out with a Vario Elementary ELIII Microcube CHNS elemental analyser. A Perkin-Elmer 603 atomic absorption spectrophotometer, with a burner control attachment and an air-acetylene flame, was used for the determination of metal ions' concentrations after extraction. The AAS metal standards, dissolved in 0.5 N nitric acid, were used to prepare standard solutions for the construction of calibration curves using 0.002 M ethylenediaminetetraacetic acid (EDTA) solution for the dilutions. The EDTA was also used to dilute the samples to prevent formation of refractory $NiSO_4$. The elements were analysed at the following specified wavelengths (nm) for minimal interferences: 232.0 (Ni^{2+}), 240.7 (Co^{2+}), 324.7 (Cu^{2+}), 213.9 (Zn^{2+}), 248.3 (Fe^{2+}, Fe^{3+}), 413.70 (Cd^{2+}), 422.7 (Ca^{2+}), 2279.5 (Mn^{2+}) and 285.2 (Mg^{2+}).

X-ray diffraction studies were performed at 200 K using a Bruker Kappa Apex II diffractometer with monochromated Mo Kα radiation (λ = 0.71073 Å). The crystal structures were solved by direct methods using SHELXTL [5] and refined with SHELXL [6]. Carbon-bound hydrogen atoms were placed in calculated positions and refined riding. The water hydrogen atoms were located on the difference map and refined riding with the bond angles and lengths restrained. The nitrogen-bound hydrogens were located on the difference map and allowed to refine freely. All non-hydrogen atoms were refined anisotropically. Diagrams and publication material were generated using PLATON [7], and ORTEP-3 [8]. The protonation and formation constants were determined by potentiometric acid-base titrations in 10% ethanol in water using the Metrohm 794 Titrino equipped with a Metrohm LL Ecotrode. This method has been presented by us previously [9], but the only difference is the use of 0.10 M sodium perchlorate as the ionic medium. The concentration stability constants (β_{pqr}) were calculated using the computer program HYPERQUAD [10]. The pH measurements for extraction studies were performed on a Metrohm 827 pH meter using a combination electrode with 3 M KCl as an electrolyte. The Labcon microprocessor-controlled orbital platform shaker model SPO-MP 15 was used for contacting the two phases of extraction. The melting points of the solid complexes were determined with the electrothermal IA 9000 digital measuring point apparatus.

2.3. Experimental Section

2.3.1. Preparation of 2,2'-Biimidazole

The ligand 2,2'-biimidazole ($BIIMH_2$) was prepared according to a method reported in the literature [11]. Yield = 17%. M.p. = 348–350 °C. Anal. Calcd for $C_8H_6N_3$ (%): C, 53.72; H, 4.51; N, 41.77. Found: C, 53.67; H, 4.70; N, 41.32. ^{1}H NMR (CDCl$_3$) δ (ppm): 7.13 (4H, s, CH). IR (v_{max}/cm^{-1}): 1693, v(C=N$_{im}$); 3327, v(N-H).

2.3.2. Preparation of 2,2'-Alkylbiimidazoles

The alkylated derivatives of $BIIMH_2$ (extractants) were also prepared according to a method found in the literature [12].

1-Heptyl-2,2'-Biimidazole (HBIIMH)

Yield = 51%. Anal. Calcd for $C_8H_6N_3$ (%): C, 67.21; H, 8.68; N, 24.12. Found: C, 67.33; H, 8.58; N, 23.84. ^{1}H NMR (400 MHz, CDCl$_3$) δ (ppm): δ 7.30 (2H, s, CH), 7.01 (2H, s, CH), 4.39 (2H, m, CH$_2$), 1.69 (2H, m, CH$_2$), 1.16 (6H, m, CH$_2$), 0.81 (3H, t, CH$_3$). IR (v_{max}/cm^{-1}): 1691, v(C=N$_{im}$); 3227, v(N-H).

1-Octyl-2,2'-Biimidazole (OBIIMH)

Yield = 49%. Anal. Calcd for $C_8H_6N_3$ (%): C, 68.26; H, 9.00; N, 22.74. Found: C, 68.30; H, 8.89; N, 21.99. ^{1}H NMR (400 MHz, CDCl$_3$) δ (ppm): δ 7.24 (2H, s, CH), 6.99 (2H, s, CH), 4.39 (2H, m, CH$_2$), 1.59 (2H, m, CH$_2$), 1.14 (6H, m, CH$_2$), 0.80 (3H, t, CH$_3$). IR (v_{max}/cm^{-1}): 1695, v(C=N$_{im}$); 3235, v(N-H).

1-Decyl-2,2′-Biimidazole (DBIIMH)

Yield = 45%. Anal. Calcd for $C_8H_6N_3$ (%): C, 70.03; H, 9.55; N, 20.42. Found: C, 70.08; H, 9.35; N, 19.95. ^1H NMR (400 MHz, CDCl$_3$) δ (ppm): δ 7.59 (2H, s, CH), 7.00 (2H, s, CH), 4.39 (2H, m, CH$_2$), 1.58 (2H, m, CH$_2$), 1.20 (6H, m, CH$_2$), 0.82 (3H, t, CH$_3$). IR (ν_{max}/cm^{-1}): 1699, ν(C=N$_{im}$); 3233, ν(N-H).

1,1′-Bis-heptyl-2,2′-biimidazole (H$_2$BIIM)

Yield = 50%. Anal. Calcd for $C_8H_6N_3$ (%): C, 72.68; H, 10.37; N, 16.95. Found: C, 73.89; H, 10.84; N, 16.54. ^1H NMR (400 MHz, CDCl$_3$) δ (ppm): δ 7.29 (2H, s, CH), 7.01 (2H, s, CH), 4.39 (2H, m, CH$_2$), 1.59 (2H, m, CH$_2$), 1.15 (6H, m, CH$_2$), 0.79 (3H, t, CH$_3$). IR (ν_{max}/cm^{-1}): 1684, ν(C=N$_{im}$); 3327, ν(N-H).

1,1′-Bis-octyl-2,2′-biimidazole (O$_2$BIIM)

Yield = 47%. Anal. Calcd for $C_8H_6N_3$ (%): C, 73.69; H, 10.68; N, 15.63. Found: C, 73.98; H, 10.89; N, 15.05. ^1H NMR (400 MHz, CDCl$_3$) δ (ppm): δ 7.29 (2H, s, CH), 7.00 (2H, s, CH), 4.38 (2H, m, CH$_2$), 1.57 (2H, m, CH$_2$), 1.15 (6H, m, CH$_2$), 0.82 (3H, t, CH$_3$). IR (ν_{max}/cm^{-1}): 1689, ν(C=N$_{im}$); 3335, ν(N-H).

1,1′-Bis-decyl-2,2′-biimidazole (D$_2$BIIM)

Yield = 48%. Anal. Calcd for $C_8H_6N_3$ (%): C, 75.31; H, 11.18; N, 13.51. Found: C, 76.00; H, 11.65; N, 13.24. ^1H NMR (400 MHz, CDCl$_3$) δ (ppm): δ 7.30 (2H, s, CH), 7.01 (2H, s, CH), 4.41 (2H, m, CH$_2$), 1.76 (2H, m, CH$_2$), 1.24 (6H, m, CH$_2$), 0.84 (3H, t, CH$_3$). IR (ν_{max}/cm^{-1}): 1699, ν(C=N$_{im}$); 3333, ν(N-H).

2.3.3. Extraction Method

All the solvent-extraction experiments were carried out in a temperature-controlled laboratory at 25 (±1) °C. Equal volumes (5 mL) of 0.001 M metal ion solution (aqueous layer) and organic layer (contains the extractant, 2-octanol, shellsol 2325 and DNNDSA) were pipetted into 50 mL conical separating funnels. They were shaken with an automated orbital platform shaker for 30 min at an optimized speed of 200 rpm. A minimum period of 24 h was observed before harvesting the raffinates. The raffinates were filtered through a 33 mm millex-HV Millipore of 0.45 µm and diluted appropriately for analysis by AAS. The percentage extractions (%E) of the metal ions were calculated from the concentrations of the metal ions in the aqueous phase using the equation below:

$$\%E = \left(\frac{Ci - Cs}{Ci}\right) \times 100 \quad (1)$$

where Ci is the initial solution concentration (mg/L) and Cs is the solution concentration after extraction. The extraction efficiencies were investigated as a function of pH, and all the extraction curves were plotted with the SigmaPlot 11.0 program.

2.3.4. Syntheses of Metal Complexes

Sulfonate Salts

The metal sulfonate salts were prepared by mixing 1:1 equimolar solution of toluene-4-sulfonic acid (RSO$_3$H) with potassium hydroxide in absolute ethanol to produce the potassium toluene-4-sulfonate salt. The potassium toluene-4-sulfonate salt was added to M(ClO$_4$)$_2$·6H$_2$O (M = Ni^{2+}, Co^{2+}, Cu^{2+} and Zn^{2+}) in absolute ethanol. The potassium perchlorate salt was removed by centrifugation and filtered. The solution was concentrated and allowed to stand at room temperature to obtain the metal sulfonate salts.

Ni(RSO$_3$)$_2$·6H$_2$O: Color: light green. Yield = 77%. M.p. = 244–246 °C. Anal. Calcd for $C_{14}H_{26}CoO_{12}S_2$ (%): C, 33.02; H, 5.15; S, 12.59. Found: C, 32.99; H, 5.01; S, 12.49. IR (cm^{-1}): 1000–1250, ν_3(SO$_3$).

Co(RSO$_3$)$_2$·6H$_2$O: Color: pale mauve. Yield = 76%. M.p. = 243–245 °C. Anal. Calcd for $C_{14}H_{26}CoO_{12}S_2$ (%): C, 33.01; H, 5.14; S, 12.59. Found: C, 32.94; H, 5.10; S, 12.53. IR (cm^{-1}): 1000–1250, ν_3(SO$_3$).

Cu(RSO$_3$)$_2$·6H$_2$O: Color: light blue. Yield = 78%. M.p. = 242–246 °C. Anal. Calcd C$_{14}$H$_{26}$CoO$_{12}$S$_2$ (%): C, 32.71; H, 5.10; S, 12.48. Found: C, 32.66; H, 5.06; S, 12.41. IR (cm^{-1}): 1000–1250, ν_3(SO$_3$).

Zn(RSO$_3$)$_2$·6H$_2$O: Color: white. Yield = 76%. M.p = 244–246 °C. Anal. Calcd C$_{14}$H$_{26}$CoO$_{12}$S$_2$ (%): C, 32.60; H, 5.08; S, 12.43. Found (%): C, 32.55; H, 5.03; S, 12.39. IR (cm^{-1}): 1000–1250, ν_3(SO$_3$).

Preparation of Sulfonate Complexes

The preparation of coordination complexes, [M(BIIM)$_2$(RSO$_3$)$_2$], was conducted in absolute ethanol under inert conditions. Hot ethanol solution (5 mL at 60 °C) containing 5 mmol of the ligand was added dropwise to 5 mL of the metal ion solution (1 mmol) of each metal ion. Toluene sulfonic acid (RSO$_3$H) (4 mmol) was added to dissolve the ligand. The mixture was heated at reflux overnight and precipitates were obtained, and these were filtered and washed with cold ethanol. A single crystal in the complex [Cu(BIIM)$_2$(RSO$_3$)$_2$] was obtained by slow diffusion of diethyl ether into the mother liquor in a desiccator at room temperature for about one month.

[Ni(BIIMH$_2$)$_2$(RSO$_3$)$_2$]: Color: green. Yield = 58%. M.p. = 243–246 °C. IR (cm^{-1}): 3311 ν(N-H), 1427 ν(C=N), 1332 ν(SO$_3$).

[Co(BIIMH$_2$)$_2$(RSO$_3$)$_2$]: Color: pink. Yield = 75%. M.p. = 241–244 °C. IR (cm^{-1}): 3333, ν(N-H); 1448, ν(C=N); 1321, ν(SO$_3$).

[Cu(BIIMH$_2$)$_2$(RSO$_3$)$_2$]: Color: green. Yield = 79%. M.p. = 245–247 °C. IR (cm^{-1}): 3323, ν(N-H); 1437, ν(C=N); 1322, ν(SO$_3$).

[Zn(BIIMH$_2$)$_2$](RSO$_3$)$_2$: Color: white. Yield = 65%. M.p. = 246–248 °C. IR (cm^{-1}): 1438, ν(C=N); 1343, ν(SO$_3$).

3. Results and Discussion

3.1. Synthesis and Characterization of 2,2'-Biimidazole and Extractants

The synthesis of 2,2'-biimidazole involves cyclization via a condensation reaction, and the alkylation of the ligand was achieved by a nucleophilic attack of alkylbromide by anionic imidazole to obtain the extractant. The purity of the products was investigated by microanalysis and confirmed by ^{1}H NMR.

The ^{1}H NMR spectrum of BIIMH$_2$ showed a peak at 7.13 ppm which was due to the four imidazole protons. The protons are chemically equivalent due to the C$_2$ symmetry of this compound. All ^{1}H NMR spectra of the ligand and extractants are provided in the Supplementary Materials (Figures S1–S6). The mono- and *bis*-alkylated biimidazole were successfully synthesized, and the appearance of the protons of the alkylated biimidazole at 7.01 and 7.30 ppm are in agreement with values found in the literature [11]. The appearance of peaks in the region of 0.070 to 4.5 ppm shows evidence of the connection of the alkyl chain to the imidazole nitrogen(s).

3.2. Solvent Extraction Studies

The extraction studies were carried out in a sulfate medium to define the optimal conditions for nickel(II) specificity. The conditions for the extraction of nickel(II) ions were optimized by investigating the essential concentration of the extractant, the concentration of the synergist (DNNDSA), the necessary alkyl chain substituent on imidazole and the effect of pH. Extractions required excess DNNDSA relative to quantities of dinonylnaphthalene sulfonic acid (DNNSA) used previously [3].

Figure S7 shows the effect of various mole ratios of Ni:OBIMH (1:25 to 1:40) on nickel(II) extraction. From these curves and the consequent data in Table S1, the ratio 1:30 showed a better extraction in terms of the steepness of the curve, i.e., a left-shifted curve and a slightly higher percentage extraction. For this reason, this metal to extractant molar ratio was chosen for the subsequent studies.

The involvement of DNNDSA as a synergist has been proven to be essential to this extraction method since there was a lack of extraction in its absence (Figure S8). This could be rationalized on the basis that the sulfate ions do not readily phase-transfer the cationic complexes formed in the extraction system from the aqueous to the organic layer. This is due to the high hydration energies offered by the sulfate ions [13]. Therefore, the application of DNNDSA, which is a bulky organic acid with very low pK_a values, eliminates the drawback posed by the sulfate ions. The role of the synergist (DNNDSA) is known to be that of an ion-pairing agent for the cationic metal ion complexes; therefore, the concentration of DNNDSA was investigated. Table S2 shows the extraction percentage as a function of pH. Initially low concentrations of DNNDSA were employed but this yielded low extraction efficiencies (Figure S9). The high concentrations resulted in significant extraction to the effect that 0.5 M was taken as optimal not because there was no further increase in %E with an increase in DNNDSA concentration but because the concentration used was already very high in comparison to the DNNSA used previously [3]. The use of such a high concentration could be expected to have negative effects on the selective extraction of Ni^{2+} ions.

DNNDSA can be expected to behave similarly to DNNSA in terms of its extraction behavior. In view of this, it is not expected to show any selectivity between metal ions, as was shown for DNNSA [3]. The DNNSA-only extractions do not show a separation of the extraction curves of the base metals' ions as a function of pH (Table S3). This necessitates the use of a ligand that has been carefully designed to cause separation between the metal ions by exploiting the bonding preferences concerning coordination numbers, stereochemistry and type of bonding involved. Du Preez has coined the term stereochemical "tailor-making" to describe this effect [1]. This effect has been demonstrated in the separation of nickel and cobalt despite their similar coordination chemistry [14].

The optimized conditions for the concentration of the extractant (L) and co-extractant were a 1:30 Ni:L ratio for a 0.001 M nickel solution and 0.5 M for DNNDSA. The effect of the alkyl substituent on the extraction efficiencies is presented in Figure S10. It seems, from this investigation, that the octyl group gives the best extraction, as evidenced by higher percentage extraction (Table S4) and steepness of the curve. An investigation was also carried out to understand the effect of monoalkylated or *bis*alkylation on imidazole, and the monoalkylated OBIIMH showed better extraction compared with the *bis*-alkylated O₂BIIM (Figure S11, Table S5). The better performance (steeper curve) of OBIIMH was probably due to its less bulky nature, thus causing less entanglement of the alkyl chains with the neighboring imidazole.

The extraction patterns of the other metal ions typically present in a leach concentrate, including Cu^{2+}, Fe^{2+}, Zn^{2+}, Fe^{3+}, Mn^{2+}, Mg^{2+}, Cd^{2+} and Ca^{2+}, were investigated under the conditions that were optimized for the extraction of nickel(II) ion (Figure 2, Table S6). There was no rejection of the hard ions (Fe^{3+}, Mn^{2+} and Ca^{2+}) in the pH range under investigation, which does not coincide with the suggestion on the bonding nature of the aromatic nitrogenous ligands. The position of the copper extraction curve is rather surprising owing to the relatively higher acidic character of this metal ion, which should make it more reactive at the lower pH region. It is clear from the extraction pattern in Figure 2 that there is a lack of pH-metric separation of the metal ions, and the differences in %E may be influenced by the solubility of chelates formed which affect the distribution between the two phases.

In this solvent-extraction system, the protonation, complexation and phase distribution equilibria can be used to describe the system quantitatively with respect to the distribution ratio of a metal ion (M^{n+}), and provide information on the coordination numbers involved in the extraction reaction [15]. The chelating agent (L) must distribute between the organic and aqueous phases to result in complexation in the aqueous phase, and that distribution coefficient is represented by $K_D(L)$:

$$(L)_a \rightleftharpoons (L)_o \text{ and } K_D(L) = \frac{[L]_o}{[L]_a} \qquad (2)$$

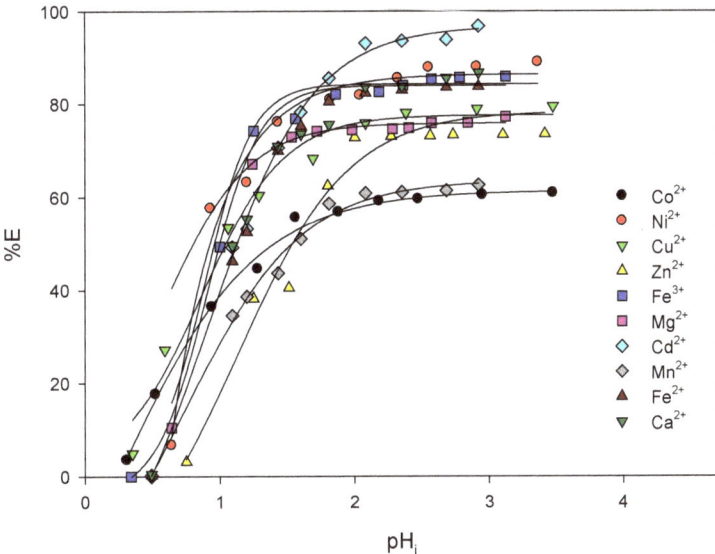

Figure 2. A plot of %E vs. initial pH of equimolar concentration (0.001 M) of Ni^{2+}, Co^{2+}, Cu^{2+}, Fe^{2+}, Zn^{2+}, Fe^{3+}, Mn^{2+}, Mg^{2+}, Cd^{2+} and Ca^{2+} extracted with OBIIMH (at M:L ratio of 1:30) and 0.5 M DNNSA in 80% 2-octanol/Shellsol 2325 from dilute sulfate medium.

However, in the aqueous phase, the following two protonation equilibria may exist depending on the pH:

$$LH_2^{2+} \rightleftharpoons H^+ + LH^+, \; K_{a1} = \frac{[H^+]_a[LH^+]_a}{[LH_2^{2+}]_a} \tag{3}$$

$$LH^+ \rightleftharpoons H^+ + L, \text{ and } K_{a2} = \frac{[H^+]_a[L]_a}{[LH^+]_a} \tag{4}$$

The metal ion chelates react with the neutral form of the ligand to form a cationic complex:

$$M^{n+} + mL \rightleftharpoons ML_m^{n+} \text{ and } K_f = \frac{[ML_m^{n+}]_a}{[M^{n+}]_a[L]_a^m} \tag{5}$$

The chelate which is ion-paired by an anion (in this case sulfonate anions represented by X^{n-}) to form an extractible species, $[ML_m]X$, distributes itself between the organic and aqueous phases:

$$\left(ML_m^{n+}\right)_a + \left(X^{n-}\right)_{o/a} \rightleftharpoons (ML_mX)_o, \text{ and } K_D(ML_m^{n+}) = \frac{[ML_mX]_o}{[ML_m^{n+}]_a} \tag{6}$$

The distribution ratio (D), defined as the ratio of the concentration of the total metal species in the organic phase to that in the aqueous (regardless of its mode), is given by Equation (7), on the assumption that the metal chelate distributes largely in the organic phase and that the metal ion does not hydrolyse in the aqueous phase.

$$D \approx \frac{[ML_mX]_o}{[M^{n+}]_a} \tag{7}$$

Substituting Equations (5) and (6), respectively, into Equation (7) yields Equation (8), depicting the formation constant and the concentration of the ligand in the aqueous phase as important parameters as well as the distribution coefficient of the chelate:

$$D = K_D(ML_m{}^{n+}) \, K_f \, [L]_a^m \tag{8}$$

Equation (8) can be transformed to Equation (9) if Equation (2) is substituted, indicating that the concentration of L in the aqueous phase is dependent on its concentration in the organic phase and that its distribution between the two phases affects the distribution ratio of the complex formed:

$$D = \frac{K_D(ML_m{}^{n+}) \, K_f}{K_D(L)^m} \, [L]_o^m \tag{9}$$

However, since the extractions are carried out at a low pH, it is necessary to consider the two protonation equilibria, respectively, because these species occur over a wide pH range, and competition of metal ions with protons for the ligand occurs early with pH due to the higher formation constants and the relatively low protonation constants (Section 3.3). Now, substituting Equations (3) and (4), respectively, into Equation (9) yields the following respective Equations (10) and (11):

$$D = K_D(ML_m{}^{n+}) \, K_f \, K_{a2}{}^m \, \frac{[LH^+]_a^m}{[H^+]_a^m} \tag{10}$$

and

$$D = K_D(ML_m{}^{n+}) \, K_f \, K_{a2}{}^m \, K_{a1}{}^m \, \frac{[LH_2{}^{2+}]_a^m}{[H^+]_a^{2m}} \tag{11}$$

Therefore, in the pH range where the monoprotonated species and a free ligand (Equation (4)) are involved, then a plot of log D vs. pH (from taking the logarithms of both sides in Equation (10)) should yield a straight line with slope m (number of ligands bonded to the metal ion M^{n+}). But in the highly acidic region where the second proton equilibrium (Equation (3)) is also active, then a plot of log D vs. pH (from Equation (11)) should yield a straight line with slope 2 m. It is therefore not surprising that the slope of the plots is steeper in the lower pH range and flattens to about 2 as the pH increases (Figure 3). This accounts for the two-stage protonation, as a higher proportion of the ligand will be monoprotonated with an increase in pH. Therefore, *bis* coordination (m ≈ 2) of 2,2'-biimidazole is supported by the extraction data.

3.3. Solution Complexation Studies

The ligand exhibits a two-stage protonation/deprotonation process and the highest log K_{a1} = 5.96 (at 25 °C) and the lowest value correspond to the protonation/deprotonation of the second imidazole group (log K_{a2} = 3.25 at 25 °C). Therefore, the loss of the first proton of the diprotonated species happens with ease and at a relatively low pH. The bidentate character in coordination was evidenced by the high formation constants that were calculated for the formation of the metal ion complexes with $BIIMH_2$ (Table 1). The overall second stability constants are of the order Cu^{2+} (10.9) > Ni^{2+} (10.7) > Zn^{2+} (10.6) > Co^{2+} (10.3). This is not strictly the same order that was observed in the extraction curves at least for Ni^{2+} and Cu^{2+} (Figure 4), but these measure values are within the range of experimental error from one another. Nonetheless, this result shows that there is no relative stability preference between $BIIMH_2$ and the base metal ions. Therefore, this extraction system is possibly governed by similar thermodynamics of complexation and results in a lack of pH-metric separation of the metal ions.

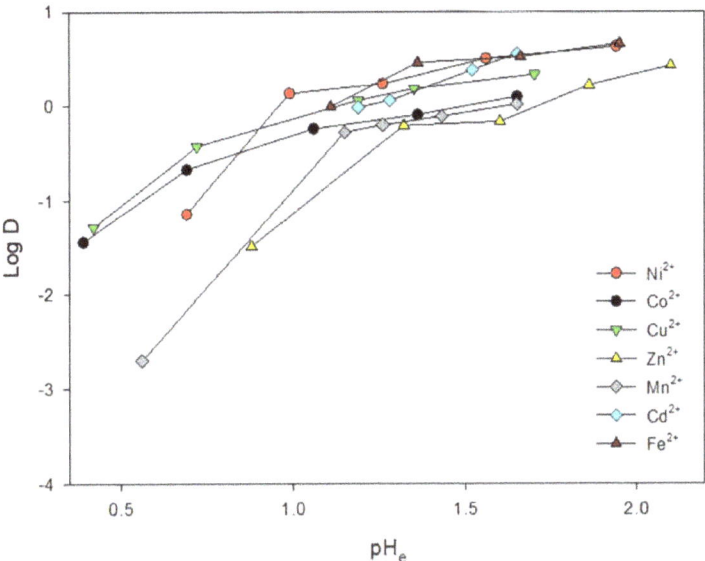

Figure 3. A plot of log D vs equilibrium pH (pH_e) for the extraction of 0.001 M M^{2+} (M = Mg^{2+}, Zn^{2+}, Mn^{2+}, Fe^{2+}, Ni^{2+}, Co^{2+} and Cu^{2+}) with OBIIMH (at M:L ratio of 1:30) and 0.5 M DNNSA in 80% 2-octanol/Shellsol 2325 from dilute sulfate medium.

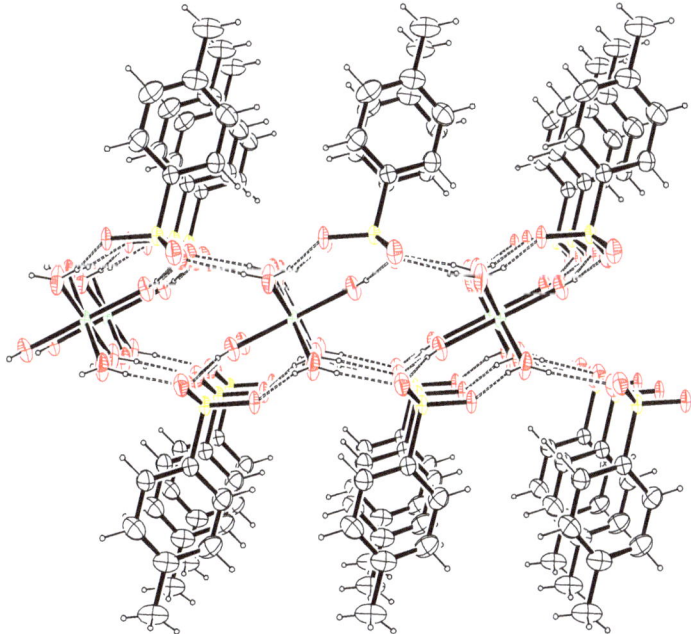

Figure 4. ORTEP diagram of $Co(RSO_3)_2 \cdot 6H_2O$ viewed normal to (1 0 0) showing the hydrogen interactions.

Table 1. Protonation and stability constants (logβ) for the interaction of BIIMH$_2$ with base metal ions as determined in 10% ethanol in water at I = 0.10 M NaClO$_4$ and 25 ($\pm$0.1) °C.

Constant	Reaction	p q r	BIIMH$_2$	Ni^{2+}	Co^{2+}	Cu^{2+}	Zn^{2+}
logβ_1	LH$^+$ = H$^+$ + L	0 1 1	5.96(5)				
logβ_2	LH$_2^{2+}$ = 2H$^+$ + L	0 1 2	9.21(5)				
logβ_{110}	M^{2+} + L = [ML]$^{2+}$	1 1 0		5.6(2)	5.3(2)	#	5.2(1)
logβ_{120}	M^{2+} + 2L = [ML$_2$]$^{2+}$	1 2 0		10.7(1)	10.3(3)	10.9(2)	10.6(1)

p, q and r refer to the coefficients of the species in the order of metal, ligand and proton. # = constant could not be calculated from current potentiometric data.

3.4. Synthesis and Characterization of Metal Complexes

A solution of toluene-4-sulfonic acid (RSO$_3$H) (10 mmol) was mixed with an equimolar amount of potassium hydroxide (10 mmol) in absolute ethanol to produce toluene-4-sulfonate salt, which was filtered and left to dry at room temperature.

$$RSO_3H + KOH \rightleftharpoons K^+ RSO_3^- + H_2O \ (R = toluene) \quad (12)$$

$$2K^+ RSO_3^- + M(ClO_4)_2 \rightleftharpoons M(RSO_3)_2 + 2KClO_4(s) \quad (13)$$

The metal sulfonates were formed from metal perchlorates, and the resulting potassium perchlorate was filtered out of solution before concentrating the metal sulfonates. Microanalaysis data supported the formation of the hexahydrate compounds M(RSO$_3$)$_2$·6H$_2$O. The sulfonate salts were then used to synthesize the 2,2'-biimidazole complexes, and these were chareacterized by melting point, FTIR, electronic spectroscopy and single crystal X-ray crystallography. The melting points for both the metal sulfonate salts and complexes are surprisingly similar (243–248 °C).

3.4.1. Spectroscopic Characterization

The characteristic υ(N-H) frequencies as well as the υ(C=N) were found in the ranges 3311–3343 cm^{-1} and 1427–1448 cm^{-1}, respectively, signifying the presence of coordinated imidazole in the complexes [16]. The presence of the sulfonate group is indicated by the υ(SO$_3$) in the range 1321–1343 cm^{-1}.

The electronic spectrum of the nickel(II) complex showed three d-d transitions at 350–400 nm, 510–650 nm and 820–1180 nm, respectively (Figure S12). These were assigned to the $^3T_{1g}(P) \leftarrow {}^3A_{2g}(F)$, $^3T_{1g}(F) \leftarrow {}^3A_{2g}(F)$ and $^3T_{2g}(F) \leftarrow {}^3A_{2g}(F)$ transitions, which are typical of an octahedral nickel(II) complex [17]. For the Co(II) complex, bands were observed at 350–385 nm, 420–580 nm and 1000–1400 nm, which may be ascribed to $^4T_{1g}(P) \leftarrow {}^4T_{1g}(F)$, $^4A_{2g}(F) \leftarrow {}^4T_{1g}(F)$ and $^4T_{2g}(F) \leftarrow {}^4T_{1g}(F)$, respectively, for octahedral symmetry [18]. The electronic spectrum of the Cu(II) complex showed two bands at 360–560 nm and 590–1000 nm. In the D$_{4h}$ symmetry, the $^2T_{2g}$ level in octahedral geometry splits into $^2E_g + {}^2B_{2g}$, while the higher ^2E_g level is unaffected. The transitions are expected to correspond to $^2B_{2g} \leftarrow {}^2E_g$ and $^2E_g \leftarrow {}^2E_g$ levels, respectively, for a distorted octahedral Cu(II) complex [19].

3.4.2. X-ray Crystallography

The ORTEP diagrams of Co(RSO$_3$)$_2$·6H$_2$O and [Cu(BIIMH$_2$)$_2$(RSO$_3$)$_2$] are illustrated in Figures 4–6, respectively. Selected crystallographic data are presented in Table 2, and selected bond lengths and angles are in Table 3. It appeared that there were traces of twinning in Co(RSO$_3$)$_2$·6H$_2$O (<2%) but attempts to take the twinning into account were unsuccessful. The CCDC deposition numbers are CCDC 2210845 and 2210846, respectively. The hexa-aqua Co(II) complex in the Co(RSO$_3$)$_2$·6H$_2$O structure is centrosymmetric, with extensive hydrogen interactions between the coordinated water molecules and the sulfonate groups (Figure 4). These interactions link adjacent complexes in a 2D network parallel to the *ab* (0 0 1) plane (Figure S13) and are presented in Table 4. The ring interactions are also presented in Table 5 and exhibit infinite stacked interactions parallel to the *a* axis. There is a slight shortening of the bonds of the axial ligands (Co-O21 = 2.046(4)) Å in the hexa-aqua complex, while the equatorial bonds have lengths of 2.075(4) Å (Figure 5), and this is perhaps due to strong supramolecular interactions that are experienced in tosylate complexes [20–22].

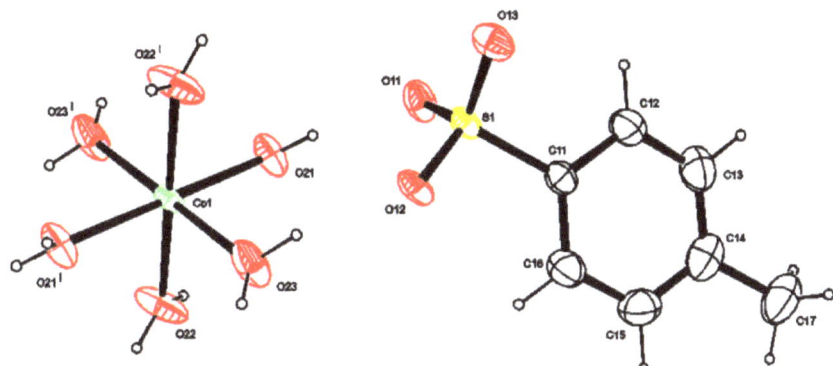

Figure 5. ORTEP diagram of Co(RSO$_3$)$_2$·6H$_2$O showing an atom numbering scheme.

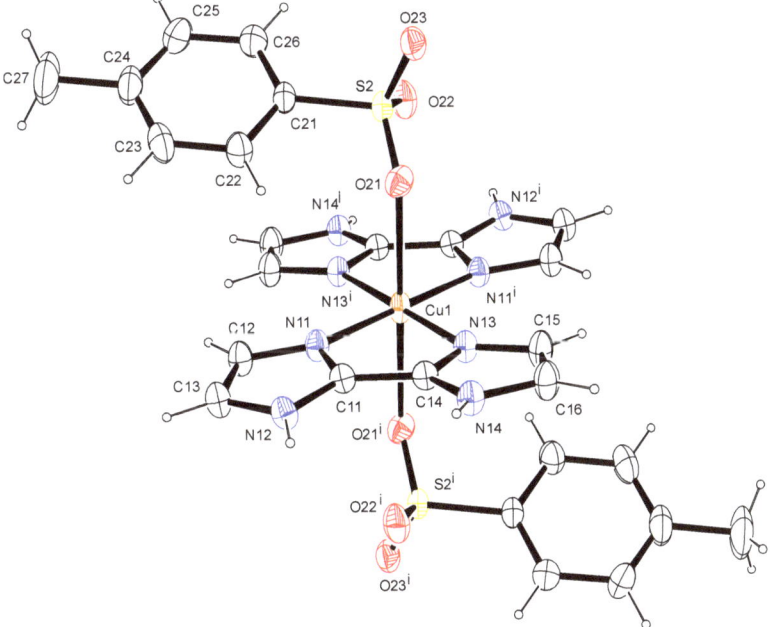

Figure 6. ORTEP diagram of [Cu(BIIM)$_2$(RSO$_3$)$_2$] with ellipsoids drawn at 50% probability. Symmetry element: (i) 1-x, 1-y, 1-z.

Table 2. Selected crystallographic data for Co(RSO$_3$)$_2$·6H$_2$O and [Cu(BIIM)$_2$(RSO$_3$)$_2$].

Compound	Co(RSO$_3$)$_2$·6H$_2$O	[Cu(BIIM)$_2$(RSO$_3$)$_2$]
Chemical formula	C$_{14}$H$_{26}$CoO$_{12}$S$_2$	C$_{26}$H$_{26}$CuN$_8$O$_6$S$_2$
Formula weight	509.40	674.24
Crystal color	pink	green
Crystal system	monoclinic	monoclinic
Space group	$P2_1/n(14)$	$P2_1/c$
Temperature (K)	200	200
Crystal size (mm^{-3})	0.06 × 0.21 × 0.37	0.06 × 0.21 × 0.37
a (Å)	6.9503(5)	12.3968(4)
b (Å)	6.2936(5)	11.7452(3)
c (Å)	25.030(2)	9.7878(3)
α (°)	90	90
β (°)	90.944(3)	91.721(2)
γ (°)	90	90
V (Å)	1094.72(15)	1424.49(7)
Z	2	2
D$_{calc}$ (g cm^3)	1.545	1.572
μ/mm^{-1}	1.031	0.970
F (000)	530	694
Theta min–max (°)	3.0, 28.4	2.4, 28.3
S	1.35	1.06
Tot., Uniq. data, R(int)	27465, 2747, 0.023	34565, 3551, 0.019
Observed data [I > 2.0σ(I)]	2638	3149
R	0.0653	0.0256
R$_w$	0.1486	0.0756

Table 3. Selected bond lengths (Å) and angles (°) for Co(RSO$_3$)$_2$·6H$_2$O and [Cu(BIIMH$_2$)$_2$(RSO$_3$)$_2$].

Bond Lengths			
Co(RSO$_3$)$_2$·6H$_2$O		[Cu(BIIMH$_2$)$_2$(RSO$_3$)$_2$]	
Co1-O21	2.046(3)	Cu1-O21	2.4302(11)
Co1-O22	2.077(4)	Cu1-N13	2.0216(12)
Co1-O23	2.076(4)	Cu1-N11'	2.0202(12)
S1-O11	1.455(4)	Cu1-N11	2.0202(12)
S1-O12	1.452(3)	Cu1-O21'	2.4302(11)
S1-O13	1.452(4)	Cu1-N13'	2.0216(12)
Bond angles			
Co(RSO$_3$)$_2$·6H$_2$O		[Cu(BIIMH$_2$)$_2$(RSO$_3$)$_2$]	
O21-Co1-O23	90.81(14)	O21-Cu1-N11	91.34(4)
O21-Co1-O23_a	89.19(14)	O21-Cu1-O21'	180.00
O21-Co1-O21_a	180.00	O21-Cu1-N13'	91.39(4)
O21-Co1-O22_a	88.90(14)	O21'-Cu1-N11	88.66(4)
O21_a-Co1-O22	88.90(14)	N11-Cu1-N13'	97.92(5)
O22-Co1-O22_a	180.00	N11'-Cu1-N13	97.92(5)
O22-Co1-O23	92.59(16)	O21'-Cu1-N11'	91.34(4)
O22-Co1-O23_a	87.41(16)	N11'-Cu1-N13'	82.08(5)
O21_a-Co1-O23	89.19((14)	O21-Cu1-N11'	88.66(4)
O22_a-Co1-O23	87.41(16)	N11-Cu1-N13	82.08(5)
O23-Co1-O23_a	180	N11-Cu1-N11'	180.00
O21_a-Co1-O22_a	91.11(14)	O21' -Cu1-N13	91.39(4)
O21_a-Co1-O23_a	90.81(14)	N13-Cu1-N13'	180.00
O22_a-Co1-O23_a	92.59(16)	O21'-Cu1-N13'	88.61(4)

Table 4. Hydrogen interactions for Co(RSO$_3$)$_2$·6H$_2$O.

D—H...A	D—H (Å)	H...A (Å)	D...A (Å)	D—H...A (°)	Symmetry
O21—H21A...O11	0.81	1.92	2.731(5)	173	
O21—H21B...O13	0.77	1.98	2.752(5)	173	1 + x, y, z
O22—H22A...O13	0.78	1.99	2.766(5)	175	1 + x, −1 + y, z
O22—H22B...O12	0.90	1.93	2.803(5)	165	1 + x, y, z
O23—H23A...O11	0.75	2.01	2.762(5)	175	x, −1 + y, z
O23—H23B...O12	0.85	1.95	2.790(5)	170	

Table 5. Short ring interactions for Co(RSO$_3$)$_2$·6H$_2$O. Cg1 is the centroid of C11–C16.

	Cg...Cg (Å)	Dihedral Angle (°)	Symmetry
Cg1...Cg1	4.924(3)	37.5(3)	−1/2−X, −1/2 + Y, 1/2−Z
Cg1...Cg1	4.988(3)	37.5(3)	1/2−X, 1/2 + Y, 1/2−Z

[Cu(BIIMH$_2$)$_2$(RSO$_3$)$_2$] is centrosymmetric, with the central copper(II) atom surrounded by two BIIMH$_2$ ligands and two oxygen atoms of the sulfonate anions (Figure 6). The geometry of the complex is distorted octahedral, with the equatorial plane formed by the four imidazole nitrogen atoms, while oxygen atoms of the sulfonate anions occupy the apical positions. The equatorial distances for the copper complex are Cu1-N11 and Cu-N11′ = 2.020(1) Å and Cu1-N13 and Cu1-N13′ = 2.0202(1) Å, while the axial Cu1-O21 and Cu1-O21′ distance is 2.430(1) Å. This is typical of a Jahn–Teller distorted copper(II) complex [15]. The majority of Cu(II) complexes are tetragonally distorted. The Cu-N and Cu-O bond lengths fall in the range normally observed for distorted octahedral copper(II) compounds [15,16,23]. The solid-state structure is in support of the observation of *bis* coordination observed in the extraction and in solution studies, as has been noticed previously [24], but this phenomenon is not always correlated due to differences in energetics of the solution vs. solid-state structures [25]. The solution/extraction studies and the electronic spectroscopic study, which suggested distorted octahedral complexes form with base metals, allowed us to conclude that the complexes are probably isostructural.

The two uncoordinated pyrrole-type nitrogens of the BIIMH$_2$ ligands both have hydrogen interactions (Table 6) with a neighboring sulfonate ligand linking adjacent complexes in an infinite 2D network parallel to the *bc* (1 0 0) plane (Figure S14). This structure also exhibits extensive ring interactions with the shortest centroid-to-centroid distance of 3.8561(9) Å (Table 7) between BIIMH$_2$ ligands forming an infinite 2D network parallel to the *bc* (1 0 0) plane.

Table 6. Hydrogen interactions for [Cu(BIIMH$_2$)$_2$(RSO$_3$)$_2$].

D—H...A	D—H (Å)	H...A (Å)	D...A (Å)	D—H...A (°)	Symmetry
N12—H12...O23	0.90(2)	1.91(2)	2.8092(17)	177.3(19)	1−x, 1/2 + y, 1/2−z
N14—H14...O22	0.884(19)	1.866(19)	2.7332(16)	166.7(17)	1−x, 1/2 + y, 1/2−z
C13—H13...O21	0.95	2.51	3.1983(18)	130	x, 3/2−y, 1/2 + z

Table 7. Short ring interactions for [Cu(BIIMH$_2$)$_2$(RSO$_3$)$_2$]. Cg1 is the centroid of N13, C14, N14, C16 and C15; Cg2 is the centroid of N11, C11, N12, C13 and C12; Cg3 is the centroid of C21–C26.

	Cg...Cg (Å)	Dihedral Angle (°)	Symmetry
Cg1...Cg2	3.8561(9)	16.87(9)	X, 3/2−Y, −1/2 + Z
Cg1...Cg3	4.5989(9)	35.97(8)	1−X, 1−Y, 1−Z

4. Conclusions

A bidentate N,N'-donor imidazole-based ligand, 2,2′-biimidazole (BIIMH$_2$), was applied as an extractant for extraction of base metal ions in an acidic sulfate medium. 1-Octyl-2-(2′-biimidazole) (OBIIMH), as a representative extractant, was applied in a solvent system, with dinonylnaphthalene disulfonic acid (DNNDSA) as a synergist. This study has established empirical evidence for the lack of separation of base metals in this system. Authors investigated the underlying chemistry and the findings from this study supported the lack of stereochemical "tailor-making" as a reason for the lack of pH-metric separation of base metals. It appears that the base metal complexes formed are isostructural with *bis* coordination of the bidentate ligand and inner sphere coordination of the sulfonate ions, and the complex formation constants are also similar, suggesting similar energetics of complexation. Of particular note is the non-innocent nature of bonding of the sulfonate ion instead of ion-pairing to form outer-sphere complexes.

Supplementary Materials: The following supporting information can be downloaded at: https://www.mdpi.com/article/10.3390/cryst13091350/s1, Figure S1: The ^{1}H NMR spectrum of 1-heptyl-2,2′-biimidazole (HBIIMH). Figure S2: The ^{1}H NMR spectrum of 1-octyl-2,2′-biimidazole (OBIIMH). Figure S3: The ^{1}H NMR spectrum of 1-decyl-2,2′-biimidazole (DBIIMH). Figure S4: The 1H NMR spectrum of 1,1′-*bis*-heptyl-2,2′-biimidazole (H$_2$BIIM). Figure S5: The ^{1}H NMR spectrum of 1,1′-*bis*-octyl-2,2′-biimidazole (O$_2$BIIM). Figure S6: The ^{1}H NMR spectrum of 1,1′-*bis*-decyl-2,2′-biimidazole (D$_2$BIIM). Figure S7: A plot of %E vs initial pH for extraction of 0.001 M nickel from dilute sulfate medium with M:L ratios 1:25, 1:30, 1:35 and 1:40 (Ni:OBIIMH) and 0.5 M DNNDSA in 80% 2-octanol/20% Shellsol 2325. Figure S8: A plot of %E vs initial pH for extraction of 0.001 M nickel from dilute sulfate medium with OBIIMH at M:L molar ratio of 1:30 in the absence of DNNDSA, and with 0.5 M DNNDSA in 80% 2-octanol/ 20% Shellsol 2325. Figure S9: A plot of %E vs. initial pH for extraction of 0.001 M nickel from dilute sulfate medium with varying concentration of DNNDSA as a synergist in 80% 2-octanol/Shellsol 2325. Figure S10: A plot of %E vs initial pH for extraction of 0.001 M nickel from dilute sulfate medium with DBIIMH, OBIIMH and HBIIMH (at Ni:L ratios of 1:30), and 0.5 M DNNSA in 80% 2-octanol/ 20% Shellsol 2325. Figure S11: A plot of %E vs initial pH for extraction of 0.001 M nickel from dilute sulfate medium with M:L ratio 1:30 (Ni:OBIIMH and Ni:O$_2$BIIM) and 0.5 M DNNDSA in 80% 2-octanol/Shellsol 2325. Figure S12: The solid reflectance spectra for nickel(II), cobalt(II) and copper(II) 2,2′-biimidazole complexes in sulfonate medium. Figure S13: ORTEP packing diagram drawn normal to (0 1 0) showing the alternating planes of complex and anion which lie parallel to the *ab* plane (0 0 1). Figure S14: Selective hydrogen interactions with ellipsoids drawn at 50 % probability. Symmetry elements: (i) 1-x, 1-y, 1-z; (ii) 1-x, 1/2+y, 1/2-z; (iii) x, 1/2-y, -1/2+z; (iv) x, 1/2-y, 1/2+z; (v) 1-x, -1/2+y, 1/2-z. Table S1: Data for %E vs initial and the equilibrium pH of 0.001 M nickel extracted from dilute sulfate medium with M:L ratios 1:25, 1:30, 1:35 and 1:40 (Ni:OBIIMH) and 0.5M DNNDSA in 80% 2-octanol/20% Shellsol 2325. Table S2: Data for %E vs. initial and the equilibrium pH for the extraction of 0.001M nickel from dilute sulfate medium with OBIIMH at a M:L molar ratio of 1:30 in the absence of DNNDSA, and with 0.5 M DNNDSA in 80% 2-octanol/20% Shellsol 2325. Table S3: Data for %E vs initial and the equilibrium pH of nickel from dilute sulfate medium with DNNDSA varying concentrations (0.01 M, 0.02 M, 0.03 M, 0.08 M, 0.1 M, 0.3 M, 0.4 M and 0.5 M) of Ni^{2+}:OBIMH (1:30) in 80% 2-octanol/20% Shellsol 2325. Table S4: Data for the extraction of nickel (0.001 M) from dilute sulfate medium with DBIIMH, OBIIMH and HBIIM at M:L ratios of 1:30, respectively, and 0.5 M DNNDSA in 80% 2-octanol/20% Shellsol 2325. Table S5: Data for %E vs initial and the equilibrium pH of 0.001 M nickel extracted from dilute sulfate medium with M:L ratio of 1:30 (Ni:OBIIMH and O$_2$BIIM) and 0.5 M DNNDSA in 80% 2-octanol/20% Shellsol 2325. Table S6: Data for %E vs initial pH of equimolar concentrations

(0.001 M) of Ni^{2+}, Co^{2+}, Cu^{2+}, Fe^{2+}, Zn^{2+}, Fe^{3+}, Mn^{2+}, Mg^{2+}, Cd^{2+} and Ca^{2+} extracted with OBIIMH at M:L ratio (1:30) and 0.5 M DNNDSA in 80% 2-octanol/20% Shellsol 2325 from sulfate medium. Table S7: Data for %E vs initial and the equilibrium pH of equimolar concentrations (0.001 M) of Ni^{2+}, Co^{2+}, Cu^{2+}, Fe^{2+}, Zn^{2+}, Fe^{3+}, Mn^{2+}, Mg^{2+}, Cd^{2+} and Ca^{2+} extracted with OPIM at M:L ratio (1:25) and 0.015 M DNNSA in 80% 2-octanol/20% Shellsol 2325 from sulfate medium. Table S8: A plot of %E vs initial and equilibrium pH in the separation of 0.001 M, Ni^{2+}, Co^{2+}, Cu^{2+}, Fe^{2+}, Zn^{2+}, Mn^{2+}, Mg^{2+} and Fe^{3+} from dilute sulfate medium with OBIMA (M:L ratios of 1:40), and 0.015 M DNNSA in 80% 2-octanol/20% Shellsol 2325.

Author Contributions: P.M.-B.—data collection, review and writing; E.C.H.—SC-XRD data collection and structure refinement and editing; Z.R.T.—Original draft proposal, conceptualization, review, writing, editing and supervision. All authors have read and agreed to the published version of the manuscript.

Funding: This research was funded by the National Research Foundation (NRF, UID 129274) and NRF-JINR (UID 120478).

Data Availability Statement: Data is provided in this article and raw data is available from researchers. CCDC 2210845 and 2210846 contains the supplementary crystallographic data for this paper. These data can be obtained free of charge from The Cambridge Crystallographic Data Centre via www.ccdc.cam.ac.uk/structures (accessed on 31 August 2023).

Acknowledgments: The authors acknowledge the National Research Foundation (NRF) of South Africa and Nelson Mandela University Research Capacity Development for financial support. We thank Mbokazi Ngayeka for reproducing some of the SigmaPlot figures.

Conflicts of Interest: Authors declare no conflict of interest.

References

1. Du Preez, J.G.H. Recent advances in amines as separating agents for metal ions. *Solvent Extr. Ion Exch.* **2000**, *18*, 679–701. [CrossRef]
2. Pearce, B.H.; Ogutu, H.F.; Luckay, R.C. Synthesis of pyrazole-based pyridine ligands and their use as extractants for nickel (II) and copper (II): Crystal structure of a copper (II)–ligand complex. *Eur. J. Inorg. Chem.* **2017**, *8*, 1189–1201. [CrossRef]
3. Okewole, A.I.; Magwa, N.P.; Tshentu, Z.R. The separation of nickel(II) from base metal ions using 1-octyl-2-(2′-pyridyl)imidazole as extractant in a highly acidic sulfate medium. *Hydrometallurgy* **2012**, *121–124*, 81–89. [CrossRef]
4. Roebuck, J.W.; Bailey, P.J.; Doidge, E.D.; Fischmann, A.J.; Healy, M.R.; Nichol, G.S.; O'Toole, N.; Pelser, M.; Sassi, T.; Sole, K.C.; et al. Strong and selective Ni(II) extractants based on synergistic mixtures of sulfonic acids and bidentate N-heterocycles. *Solvent Extr. Ion Exch.* **2018**, *36*, 437–458. [CrossRef]
5. Sheldrick, G.M. SHELXT—Integrated space-group and crystal-structure determination. *Acta Cryst.* **2015**, *A71*, 3–8. [CrossRef] [PubMed]
6. Sheldrick, G.M. Crystal structure refinement with SHELXL. *Acta Cryst.* **2015**, *C71*, 3–8.
7. Spek, A.L. Structure validation in chemical crystallography. *Acta Cryst.* **2009**, *D65*, 148–155. [CrossRef]
8. Farrugia, L.J. WinGX and ORTEP for Windows: An update. *J. Appl. Cryst.* **2012**, *45*, 849–854. [CrossRef]
9. Okewole, A.I.; Walmsley, R.S.; Valtancoli, B.; Bianchi, A.; Tshentu, Z.R. Separation of copper(II) from a highly acidic synthetic sulfate leach solution using a novel 1-octylimidazole-2-aldoxime extractant. *Solvent Extr. Ion Exch.* **2013**, *31*, 61–78. [CrossRef]
10. Gans, P.; Sabatini, A.; Vacca, A. Investigation of equilibria in solution. Determination of equilibrium constants with the HYPERQUAD suite of programs. *Talanta* **1996**, *43*, 1739–1753. [CrossRef]
11. Moleko, P. The Coordination and Extractive Chemistry of the Later 3d Transition Metal Ions with N,N'-Donor Imidazole-Based Ligands. Master's Dissertation, Nelson Mandela Metropolitan University, Gqeberha, South Africa, 2014; p. 37.
12. Haring, M. A novel route to N-substituted heterocycles. *Helv. Chim. Acta* **1959**, *42*, 1845–1850.
13. Allen, K.A. Equilibrium between didecylamine and sulphuric acid. *J. Phys. Chem.* **1956**, *60*, 943–946. [CrossRef]
14. Flett, D.S. Solvent extraction in hydrometallurgy: The role of organophosphorus extractants. *J. Organomet. Chem.* **2005**, *690*, 2426–2438. [CrossRef]
15. Magwa, N.P.; Hosten, E.; Watkins, G.M.; Tshentu, Z.R. An exploratory study of tridentate amine extractants-solvent extraction and coordination chemistry of base metals with bis((1R-benzimidazol-2-yl)methyl)amine. *Int. J. Nonferrous Metall.* **2012**, *1*, 49–58. [CrossRef]
16. Magwa, N.P.; Hosten, E.; Watkins, G.M.; Tshentu, Z.R. The coordination and extractive chemistry of the later 3d transition metals with bis((1R-benzimidazol-2-yl)methyl)sulfide. *J. Coord. Chem.* **2013**, *66*, 114–125. [CrossRef]
17. Takahashi, K.; Nishida, Y.; Kida, S. Crystal structure of copper(II) complex with N,N-bis(2-benzimidazolylmethyl)benzylamine. *Polyhedron* **1984**, *4*, 113–116. [CrossRef]

18. Pandiyan, T.; Hernandez, J.G. Geometrical isomers of bis(benzimidazol-2-ylethyl)sulfide)cobalt(II) diperchlorates: Synthesis, structure, spectra and redox behavior of pink-[Co(bbes)$_2$](ClO$_4$)$_2$ and blue-[Co(bbes)$_2$](ClO$_4$)$_2$. *Inorg. Chim. Acta.* **2004**, *357*, 2570–2578. [CrossRef]
19. Hancock, R.D.; Martell, A.E. Ligand design for selective complexation of metal ions in aqueous solution. *Chem. Rev.* **1989**, *89*, 1875–1914. [CrossRef]
20. Fewings, K.R.; Junk, P.C.; Georganopoulou, D.; Prince, P.D.; Steed, J.W. Supramolecular interactions in metal tosylate complexes. *Polyhedron* **2001**, *20*, 643–649. [CrossRef]
21. Kosnic, J.E.; McClymont, L.; Hodder, R.A.; Squattrito, P.J. Synthesis and structures of layered metal sulfonate salts. *Inorg. Chim. Acta* **1996**, *244*, 253–254. [CrossRef]
22. Cabaleiro-Martinez, S.; Castro, J.; Romero, J.; Garcia-Vazquez, J.; Sousa, A. Hexaaquacobalt(II) bis(4-toluene-sulfonate). *Acta Cryst.* **2000**, *C56*, e249–e250.
23. Rietmeijer, F.J.; Birker, P.J.M.W.L.; Gorter, S.; Reedijk, J. Copper(I) and copper(II) chelates containing imidazole and thioether groups; synthesis of the ligand 1,2-bis(benzimidazol-2′-ylmethylthio)-benzene (bbtb) and the X-ray crystal structure at 52 °C of [Cu-(bbtb)(H$_2$O)][ClO$_4$]$_2$·5EtOH. *J. Chem. Soc.* **1982**, *7*, 1191–1198. [CrossRef]
24. Wilson, A.M.; Bailey, P.J.; Tasker, P.A.; Turkington, J.R.; Grant, R.A.; Love, J.B. Solvent extraction: The coordination chemistry behind extractive metallurgy. *Chem. Soc. Rev.* **2014**, *4*, 123–134. [CrossRef]
25. de Sousa, A.S.; Fernandes, M.A.; Padayachy, K.; Marques, H.M. Amino-alcohol ligands: Synthesis and structure of *N,N*-bis(2-hydroxycyclopentyl)ethane-1,2-diamine and its salts, and an assessment of its fitness and that of related ligands for complexing metal ions. *Inorg. Chem.* **2010**, *49*, 8003–8011. [CrossRef]

Disclaimer/Publisher's Note: The statements, opinions and data contained in all publications are solely those of the individual author(s) and contributor(s) and not of MDPI and/or the editor(s). MDPI and/or the editor(s) disclaim responsibility for any injury to people or property resulting from any ideas, methods, instructions or products referred to in the content.

Article

New Polynuclear Coordination Compounds Based on 2–(Carboxyphenyl)iminodiacetate Anion: Synthesis and X-rays Crystal Structures

Sebastián Martínez [1], Carlos Kremer [1], Javier González-Platas [2] and Carolina Mendoza [1,*]

[1] Área Química Inorgánica, Departamento Estrella Campos, Facultad de Química, Universidad de la República, Montevideo 11800, Uruguay; sebamartinez@fq.edu.uy (S.M.); ckremer@fq.edu.uy (C.K.)

[2] Departamento de Física, Instituto Universitario de Estudios Avanzados en Física Atómica, Molecular y Fotónica (IUDEA), MALTA Consolider Team, Servicio de Difracción de Rayos X, Universidad de La Laguna, 38200 Tenerife, Spain; jplatas@ull.edu.es

* Correspondence: cmendoza@fq.edu.uy; Tel.: +598-29249739

Abstract: In the present work, novel polymeric copper(II) coordination compounds, namely $[Cu_2(cpida)(H_2O)_4][Cu(cpida)]\cdot 3H_2O$ (**1**) ($cpida^{3-}$ = 2-(carboxyphenyl)iminodiacetate anion) and $Na[Cu(cpida)]$ (**3**), were synthesized and characterized using infrared spectroscopy, thermogravimetric analysis, elemental analysis, and single-crystal X-ray diffraction. Compound **1** was obtained by slowly evaporating an aqueous solution of H_3cpida, copper(II) sulfate, and NaOH at room temperature. The structural characterization revealed that **1** is an ionic entity formed by the $[Cu(cpida)]^-$ anion and the $[Cu_2(cpida)(H_2O)_4]^+$ cation, both of polymeric 1D structure. Compound **3** was prepared under similar conditions from copper perchlorate and crystallized via acetone diffusion. It is a coordination polymer formed by the $[Cu(cpida)]^-$ units, and the sodium cation is present in the structure, counterbalancing the anion charge. Depending on the crystallization conditions, it was possible to obtain other solvation forms of these structures. Starting from the conditions of compound **1**, via the diffusion of ethanol, $[Cu_2(cpida)(H_2O)_4][Cu(cpida)]\cdot H_2O\cdot 1/2EtOH$ (**2**) was precipitated, while a hydrate form of compound **3** with the formula $Na[Cu(cpida)]\cdot 2H_2O$ (**4**) was obtained via methanol diffusion.

Keywords: polynuclear complexes; X-ray crystal structures; Cu(II); 2-carboxyphenyliminodiacetic acid

Citation: Martínez, S.; Kremer, C.; González-Platas, J.; Mendoza, C. New Polynuclear Coordination Compounds Based on 2-(Carboxyphenyl)iminodiacetate Anion: Synthesis and X-rays Crystal Structures. *Crystals* **2023**, *13*, 1669. https://doi.org/10.3390/cryst13121669

Academic Editor: Yael Diskin-Posner

Received: 31 October 2023
Revised: 29 November 2023
Accepted: 29 November 2023
Published: 9 December 2023

Copyright: © 2023 by the authors. Licensee MDPI, Basel, Switzerland. This article is an open access article distributed under the terms and conditions of the Creative Commons Attribution (CC BY) license (https://creativecommons.org/licenses/by/4.0/).

1. Introduction

Over the past few decades, there has been substantial development in the research on polynuclear coordination compounds and coordination polymers, which is particularly fueled by their potential applications as materials for gas storage, catalysis, environmental remediation, drug delivery, and in magnetochemistry or semiochemistry [1–13].

According to the IUPAC recommendations, coordination polymers are "coordination compounds with repeating coordination entities extending in 1, 2, or 3 dimensions" [14,15]. In the structures of these compounds, there are inorganic nodes (metal-containing clusters or metal ions) connected by organic divergent molecules (called linkers or connectors) [1,16,17]. Rigid ligands with a high degree of directionality were originally used as linkers in order to obtain specific crystal motifs [18]. However, the employment of more flexible ligands in the development of coordination polymers has lately been explored by numerous research groups [19–33]. The flexibility arises from the rotation of single bonds in carbon chains [22], allowing for various conformations that can result in distinct final structures [24].

The employment of flexible ligands as linkers in coordination polymers has encountered some hurdles when compared with the use of less flexible or rigid connectors [24]. This is due to the fact that the process of self-assembling is dependent on the reaction conditions, particularly the solution pH value, the type of metal cation, the concentrations of the different reactants, and the solvent employed. What is more, flexible polydentate ligands are also able to form mononuclear coordination compounds in competitive processes or adopt different conformations that are difficult to control.

Flexible molecules containing carboxylic acid groups are often chosen for the assembly of polynuclear coordination compounds. This preference is supported by the fact that this type of ligand can coordinate in different ways due to the possibility of fully or partially deprotonating the carboxylate groups, resulting in different topologies of the final structure [34]. These flexible molecules can act as bridging bidentate or tridentate ligands or chelating agents and even coordinate in a unidentate mode [35]. They are also good hydrogen bond acceptors capable of assembling supramolecular structures [36].

For several years now, we have focused our work on the development of polynuclear coordination compounds using polytopic flexible ligands. When working with aqueous solutions at mild acidic pH values, we reported the obtention of four capsule-like trinuclear copper(II) complexes assembled with nitrilotripropionic acid (H_3ntp) [37,38] and three new metal polynuclear compounds (containing nickel(II) or copper (II)) with N-benzyliminodipropionic acid (H_2bzlidp) [39]. In the case of the ntp complexes, the anion of copper salt had a significant influence on the obtained structure, defining the connectivity between the capsules [37]. When bzlidp^{2-} was employed as a ligand, it was possible to obtain Cu(II) complexes whose structures were significantly different (dinuclear complex vs. 2D layered structure) depending on the starting conditions [39].

We then turned our attention to the tricarboxylic acid compound 2-(carboxyphenyl) iminodiacetic acid (H_3cpida, Figure 1) as a connector of the metal ions. This is a multidentate ligand with the ability to form five-membered chelate rings. It presents flexible iminodiacetate groups and an aromatic ring that can initiate π–π interactions to stabilize the crystal packing.

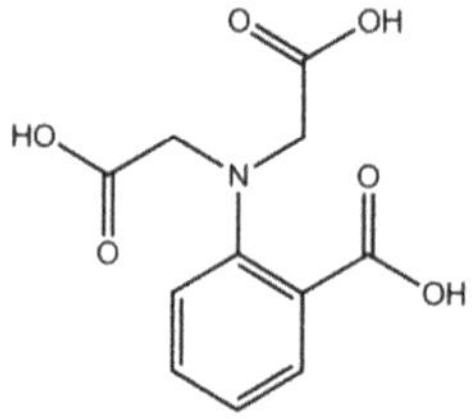

Figure 1. Structure of 2-carboxyphenyliminodiacetic acid.

This ligand has already been employed in the preparation of mononuclear [40–43] and trinuclear coordination compounds [44] and high-nuclearity metal aggregates [45–49]. However, the employment of this tricarboxylic molecule to connect metal centers in a 1D structure has been scarcely studied [44,50,51]. Bandyopadhyay and collaborators reported two polymeric copper (II) compounds, {[Cu(Hcpida)]·H_2O}$_n$ and {[Cu(Hcpida)(H_2O)]·H_2O}$_n$ [50]. In the first compound, the copper centers are pentacoordinated, showing a square pyramid geometry with the Hcpida^{2-} anion bridging the metal ions through the carboxylate groups attached to the phenyl ring. In this complex, one of the carboxylate groups remains protonated, stabilizing the chains through hydrogen bonds. In the second compound, the Cu(II) center is hexacoordinated, presenting a distorted octahedral geometry. The protonated carboxylate group is connected to the coordinated water molecule of a [Cu(Hcpida)(H_2O)] unit of an adjacent chain through a hydrogen bond. In this way, cylindrical channels are formed in the structure that hosts the water crystallization molecule. Yong and coworkers described a Cd(II) coordination polymer, [Cd$_3$(cpida)$_2$(H_2O)]·H_2O, which contains three crystallographically independent Cd(II) centers and forms infinite helical chains [51]. Heteroleptic Cu(II) chain-like structures can also be found in the literature. Ma and coworkers characterized a Cu(II) coordination polymer containing the cpida^{3-} anion and 1,10-phenanthroline (phen) with a {[Cu$_2$(cpida)(phen)(NO$_3$)]·2H_2O}$_n$ formula. The complex contains two five-coordinated copper(II) centers with a square pyramid geometry. Two cpida^{3-} anions bridge three copper ions, forming a trinuclear triangular cluster that, in turn, is connected by their vertices, forming a 1D structure [44]. Polymeric coordination compounds based on the meta and para isomers of the 2-(carboxyphenyl)iminodiacetate anion have also been reported [35,52–54].

We decided to further explore the potential of H_3cpida to form polynuclear systems, in conditions analogous to those employed in the preparation of the H_2bzlidp or H_3ntp compounds mentioned above. Copper(II) was used as a metal ion and we report the preparation of four new copper(II) polynuclear compounds, [Cu$_2$(cpida)(H$_2$O)$_4$][Cu(cpida)]·3H$_2$O (**1**), [Cu$_2$(cpida)(H$_2$O)$_4$][Cu(cpida)]·H$_2$O·1/2EtOH (**2**), Na[Cu(cpida)] (**3**), and Na[Cu(cpida)]·2H$_2$O (**4**), obtained under different crystallization conditions.

2. Materials and Methods

2.1. General Information

All common laboratory chemicals were reagent grade, acquired from commercial suppliers, and employed without additional purification. Elemental analysis (C, H, N, S) was carried out on a Flash 2000 instrument (Thermo Fisher Scientific, USA). The infrared spectra, as KBr pellets, were obtained on an FTIR Shimadzu IR-Prestige-21 spectrophotometer (Shimadzu, Kyoto, Japan) in the 4000–400 cm^{-1} range. Thermogravimetric analyses (TGAs) were executed in a Shimadzu TGA-50 instrument with a TA 50 l interface (Shimadzu, Kyoto, Japan), using platinum cells. TGAs were recorded under a nitrogen stream (50 mL min^{-1}) at 1.0 °C min^{-1} up to 300 °C and 30 °C min^{-1} from 300 to 700 °C.

CAUTION! Perchlorate salts of metal coordination compounds containing organic ligands have the potential to be explosive. Only small quantities of material should be prepared and it should be handled with caution.

2.2. Synthesis of the Complexes

2.2.1. Synthesis of [Cu$_2$(cpida)(H$_2$O)$_4$][Cu(cpida)]·3H$_2$O (**1**)

H_3cpida (0.167 g, 0.66 mmol) was dissolved in distilled water (12 mL) and copper sulfate (0.159 g, 1.00 mmol) was added. The solution was magnetically stirred at room temperature and adjusted to pH 4.1 with 5 mol L^{-1} NaOH solution. The blue solution was filtered and the solvent was slowly evaporated at room temperature. After four months, **1** precipitated as green crystals. The crystals were filtered through paper and washed with cold distilled water (0 °C). The solid material was dried in open air, and 0.084 g of the product was obtained. The yield was 31% (based on copper(II)). Anal. Calc. for C$_{22}$H$_{30}$Cu$_3$N$_2$O$_{19}$; C, 32.34; H, 3.70; N, 3.43. Found: C, 31.90; H, 3.45; N 3.39%. Selected IR data (ν_{max}/cm^{-1}): 3447, 3178, 2968 (sh), 2929 (sh), 1599, 1549, 1481, 1454, 1410, 1306, 1259, 1175, 1117, 1084, 1032, 1001, 990, 957, 932, 880, 812, 763, 723, 710, 646,594, 527, 446. TGA (30–700 °C, N$_2$), % weight loss: 11.29 (left limit 19 °C–right limit 63 °C), 4.95 (left limit 63 °C–right limit 139 °C), 38.95 (left limit 170 °C –right limit 328 °C), 7.26% (left limit 328 °C–right limit 700 °C).

2.2.2. Synthesis of [Cu$_2$(cpida)(H$_2$O)$_4$][Cu(cpida)]·H$_2$O·1/2EtOH (**2**)

H_3cpida (0.253 g, 1.0 mmol) was dissolved in distilled water (6 mL) and the pH was set to 7 through addition of 5 mol L^{-1} NaOH. This solution was added to a 2 mL aqueous solution of copper trifluoromethanesulfonate (0.543 g, 1.5 mmol). The solution was stirred at room temperature and its pH adjusted to 5.2 with aqueous 5 mol L^{-1} NaOH solution. This solution was subsequently divided into two portions of 4 mL each, which were set to crystallize either by diffusion of ethanol or methanol vapors. After 21 d, green crystals were obtained from the first (ethanol vapors). The crystals were filtered through paper and washed with cold distilled water (0 °C). After allowing the crystals to dry in open air, 0.070 g of **2** was obtained. The yield was 35% (based on copper(II)). Anal. Calc. for C$_{46}$H$_{58}$Cu$_6$N$_4$O$_{35}$: C, 34.35; H, 3.64; N, 3.48. Found: C, 33.54; H, 3.61; N 3.34%. Selected IR data (ν_{max}/cm^{-1}): 3447, 3145 (sh), 2972, 2928, 2855, 1609, 1564, 1547, 1479, 1454, 1408, 1373, 1304, 1258, 1173, 1121, 1086, 1040, 999, 957, 934, 876, 808,766, 721, 712, 646, 590, 525.

Blue needle-shaped crystals of a different compositions were obtained from the diffusion of methanol (see below).

2.2.3. Synthesis of Na[Cu(cpida)] (3)

H$_3$cpida (0.253 g, 1.0 mmol) was dissolved in 6 mL of distilled water and the pH value of the resulting solution was set to 7 by adding NaOH (5 mol L^{-1} solution). This ligand solution was added to a 2 mL aqueous solution of copper perchlorate hexahydrate (0.556 g, 1.5 mmol). The solution was stirred at room temperature and its pH was set to 5.2, again with 5 mol L^{-1} NaOH solution. Blue needle-shaped crystals were obtained after 60 days, produced from 2 mL of this solution via diffusion of acetone vapors. These crystals were filtered, washed with cold distilled water (0 °C), and allowed to dry in open air. Following this procedure, 0.034 g of the product was obtained. The yield was 40% (based on H$_3$cpida). Anal. Calc. for C$_{11}$H$_8$CuNNaO$_6$; C, 39.24; H, 2.39; N, 4.16. Found: % C, 38.82; H, 2.42; N 4.06%. Selected IR data (ν_{max}/cm^{-1}): 3441, 2979, 2939, 1674, 1638, 1616, 1483, 1450, 1396, 1385, 1339, 1219, 1171, 1152, 1103, 1061, 1045, 961, 941, 908, 881, 853, 777, 752, 669, 617, 575, 550, 523, 442.

2.2.4. Synthesis of Na[Cu(cpida)]·2H$_2$O (4)

H$_3$cpida (0.253 g, 1.0 mmol) was dissolved in 6 mL of distilled water and the pH was set to 7 with the addition of NaOH (5 mol L^{-1} solution), drop by drop. This solution was added to 2 mL of aqueous copper perchlorate hexahydrate solution (0.556 g, 1.5 mmol). The mixture was magnetically stirred at room temperature and the pH was set to 5.2 with aqueous NaOH solution. Blue needle-shaped crystals of compound **4** were obtained after 50 days, produced from 2 mL of the resulting solution via diffusion of methanol vapors. The crystals were separated by filtration, washed with cold water (0 °C), and dried in open air. Following this procedure, 0.035 g of the product was obtained. The yield was 38% (based on H$_3$cpida) Anal. Calc. for C$_{11}$H$_{12}$CuNNaO$_8$, C, 35.44; H, 3.25; N, 3.76. Found: C, 35.05; H, 3.12; N, 3.54%. Selected IR data (ν_{max}/cm^{-1}): 3423, 2970, 2932, 1657, 1634, 1481, 1447, 1402, 1369, 1346, 1317, 1228, 1182, 1163, 1148, 1099, 1063, 1049, 982, 959, 941, 910, 880, 853, 766, 752, 708, 671, 652, 617, 578, 552, 529, 492, 447. TGA (30–700 °C, N$_2$) % weight loss: 8.00 (left limit 19 °C–right limit 120 °C), 34.66 (left limit 100 °C–right limit 305 °C), 20.96 (left limit 305 °C–right limit 700 °C).

Compound **4** could also be obtained from copper trifluoromethanesulfonate, in a similar manner as compound **2** but using methanol for the diffusion (0.0315 g).

2.3. X-ray Crystallography

The X-ray diffraction data for all compounds were collected with a Rigaku SuperNOVA diffractometer with microfocus X-ray using Cu/Mo Kα radiation (λ = 1.54184/0.71073 Å). CrysAlisPro 1.171.39.46 [55] software was employed to collect, index, scale, and apply analytical absorption correction based on the faces of the crystal. The SHELXT program was used to solve the structure with a dual-space algorithm [56]. Fourier recycling and least-squares refinement were used for the model completion with SHELXL-2019 [57]. All non-hydrogen atoms were refined anisotropically, and all hydrogen atoms were placed in geometrically suitable positions and refined through riding with the isotropic thermal parameter related to the equivalent isotropic thermal parameter of the parent atom. The hydrogen atoms were geometrically positioned with C-H = 0.93 Å and Uiso(H) = 1.2 Ueq(C). The geometrical analysis of interactions in the structure was performed with the Olex2 [58] program (version 1.5).

The crystal data, collection procedures, and refinements results are summarized in Table 1.

Table 1. Crystallographic data for compounds **1–4**.

Compound	1	2	3	4
CCDC number	2304883	2304881	2304880	2304882
Empirical formula	$C_{22}H_{30}Cu_3N_2O_{19}$	$C_{23}H_{29}Cu_3N_2O_{17.5}$	$C_{11}H_8CuNNaO_6$	$C_{11}H_{12}CuNNaO_8$
Formula weight	817.10	804.10	336.71	372.75
Temperature (K)	293(2)	293(2)	150(2)	150(2)
Wavelength	0.71073 Å	1.54184 Å	1.54184 Å	1.54184 Å
Crystal system	Monoclinic	Monoclinic	Orthorhombic	Orthorhombic
Space group	P $2_1/n$	P $2_1/n$	Pbca	Iba2
a (Å)	18.0990 (7)	18.1311 (4)	14.4270 (8)	28.1538 (16)
b (Å)	9.7523 (3)	9.7972 (2)	6.4340 (4)	14.5726 (7)
c (Å)	18.2152 (6)	18.1195 (5)	23.3780 (16)	6.2930 (3)
α (°)	90	90	90	90
β (°)	93.438 (3)	94.229(2)	90	90
γ (°)	90	90	90	90
Volume (Å^3)	3209.32 (19)	3209.88 (13)	2170.0 (2)	2581.9 (2)
Z	4	4	8	8
D_{calc} (g·cm^{-3})	1.691	1.664	2.061	1.918
Absorption coefficient (mm^{-1})	2.053	3.011	3.540	3.164
F(000)	1660	1632	1352	1512
Crystal dimensions (mm)	0.22 × 0.17 × 0.11	0.43 × 0.30 × 0.22	0.19 × 0.07 × 0.03	0.33 × 0.11 × 0.05
Theta range for data collection (°)	1.636 to 26.372	5.135 to 66.585	4.869 to 67.033	5.607 to 66.556
Index ranges	$-14 \leq h \leq 22$, $-12 \leq k \leq 12$, $-21 \leq l \leq 22$	$-21 \leq h \leq 19$, $-11 \leq k \leq 8$, $-18 \leq l \leq 21$	$-17 \leq h \leq 16$, $-7 \leq k \leq 4$, $-27 \leq l \leq 12$	$-33 \leq h \leq 27$, $-12 \leq k \leq 17$, $-7 \leq l \leq 4$
Reflections collected	13,135	12,308	4666	2980
Independent reflections, R(int)	6552, 0.0367	5452, 0.0163	1927, 0.0284	1599, 0.0206
Data/restraints/parameters	6552/3/437	5452/0/421	1927/0/181	1599/1/181
Goodness of fit on F^2	1.022	1.091	1.173	1.081
Final R indices [I > 2sigma(I)][a,b]	$R_1 = 0.0485$, $wR_2 = 0.1251$	$R_1 = 0.0431$, $wR_2 = 0.1338$	$R_1 = 0.0571$, $wR_2 = 0.1406$	$R_1 = 0.0255$, $wR_2 = 0.0679$
R indices (all data)[a,b]	$R_1 = 0.0649$, $wR_2 = 0.1375$	$R_1 = 0.0451$, $wR_2 = 0.1353$	$R_1 = 0.0643$, $wR_2 = 0.1438$	$R_1 = 0.0259$, $wR_2 = 0.0682$
Largest diff. peak and hole (e.Å^{-3})	0.90 / −0.54	1.65 / −0.88	0.81 / −0.82	0.37 / −0.41

[a] $R_1 = \Sigma ||F_0| - |F_c|| / \Sigma |F_c|$, [b] $wR_2 = (\Sigma[w(F_0^2 - F_c^2)^2] / \Sigma[w(F_0^2)^2])^{1/2}$.

3. Results and Discussion

3.1. Synthesis of the Complexes

The coordination compounds were prepared by directly mixing aqueous solutions of the corresponding copper salts and H$_3$cpida, setting the pH value in the 4.0–5.2 range, and using a 1.5 copper-to-ligand ratio (Scheme 1).

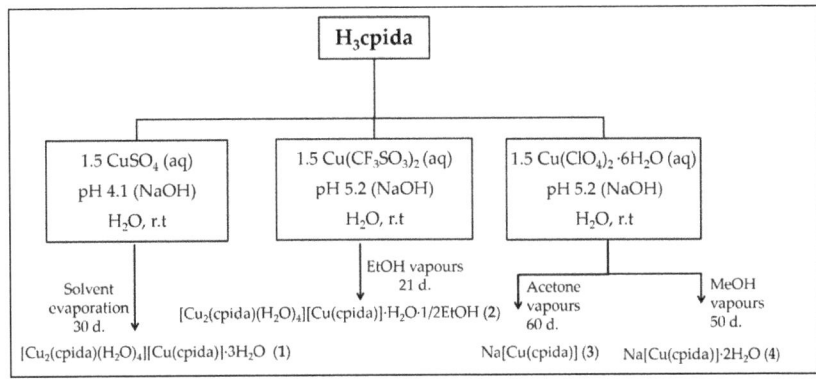

Scheme 1. Synthesis of the complexes **1–4**.

The crystals were characterized via infrared spectroscopy and elemental analysis. IR spectra of compounds **1, 2,** and **4** exhibit a broad band in the 3200–3600 cm^{-1} region. The presence of this band is indicative of the existence of water molecules in the crystal structures. The small absorption peaks that appeared in the 3000–2850 cm^{-1} region for the four complexes correspond to the ν_{C-H} vibration modes of the –CH$_2$– groups in the alkyl chains on the cpida^{3-} anion (Figures S1–S4 Supplementary Material). The absence of the distinctive band of ν_{as}(COOH) at *ca.* 1700 cm^{-1} indicates that the carboxylate groups of the ligand are fully deprotonated in these compounds. For complexes **1** and **2**, the ν_{as}(COO$^-$) vibrations appeared at ca. 1600 cm^{-1}, whereas those of ν_s(COO$^-$) occurred at ca. 1410 cm^{-1}. In the case of compounds **3** and **4**, the bands around 1635 cm^{-1} can be assigned to the asymmetric stretching of the COO$^-$ of the complexes while those that appeared in the proximity of 1400 cm^{-1} correspond to ν_s(COO$^-$). In all cases, the elemental analysis is consistent with the proposed empirical formula.

The TGA curve for compound **1** (Figure S5, Supplementary Material) shows, below 140 °C, a dehydration process that occurs in two steps, with a weight loss of 16.2%. The dehydration process is consistent with the loss of seven water molecules, comprising both lattice water molecules and coordinated ones. This is followed by a 46% weight loss step that involves the decomposition of the organic ligand. This is similar to what has previously been observed in the TG analysis of complexes containing the cpida ligand [43,44,50,51,53]. In the case of compound **4**, the TGA curve (Figure S6, Supplementary Material) shows a weight loss of 8% below 120 °C, consistent with a two-molecule dehydration process, and a two-step decomposition process with a weight loss of circa 56%, corresponding to the decomposition of the organic ligand. Single crystals for the complexes were obtained through slow evaporation of the solvent (**1**), and slow diffusion of ethanol (**2**), acetone (**3**), and methanol (**4**), at room temperature. Crystallographic data of compounds **1** to **4** are summarized in Table 1.

*3.2. Crystal Structure of [Cu$_2$(cpida)(H$_2$O)$_4$][Cu(cpida)]·3H$_2$O (**1**) and of [[Cu$_2$(cpida)(H$_2$O)$_4$][Cu(cpida)]·H$_2$O·1/2EtOH] (**2**)*

Compound **1** crystallizes in the monoclinic P 2$_1$/n space group (Table 1). Selected bond lengths and angles for **1** are presented in Table 2 and in Table S1 (Supporting Information). The asymmetric unit of [Cu$_2$(cpida)(H$_2$O)$_4$][Cu(cpida)] is shown in Figure 2a. Compound **1** is a salt formed by the [Cu(cpida)]$^-$ anion and the [Cu$_2$(cpida)(H$_2$O)$_4$]$^+$ cation, both of them being polymeric, in a type of structure that is quite common with this kind of iminocarboxylic ligand [39,50]. The structure also presents three lattice water molecules. The anion is formed by [Cu(cpida)] units, connected by carboxylate oxygen atoms. Each unit contains a Cu(II) ion coordinated to a cpida^{3-} anion, which acts as a tetradentate ligand. In order to assess the coordination geometry of Cu2, which shows five ligated atoms, the τ factor was determined. This is a geometrical parameter proposed by Addison and coworkers [59] to classify five coordinated structures in square pyramidal (τ factor = 0) or trigonal bipyramids (τ factor = 1). According to this parameter, the coordination polyhedron of the Cu2 atom can be depicted as a slightly distorted square pyramid (τ factor = 0.33), the donor atoms being the nitrogen atom N2, and three oxygen atoms of each carboxylate of the cpida^{3-} anion (O8, O9, and O12, Figure 2b). The coordination sphere is completed by the O-carboxylate of a neighboring [Cu(cpida)] unit. Therefore, two of the carboxylate groups are coordinating in a monodentate mode, while the third is acting as bridging bidentate, in a *syn-anti* conformation and with a Cu–Cu separation of 5.25(7) Å. Cu(II) ida or substituted ida complexes have shown this kind of connection [44,60–62] as opposed to the one previously reported [50] where the bridging carboxylate group was the one attached to the phenyl ring.

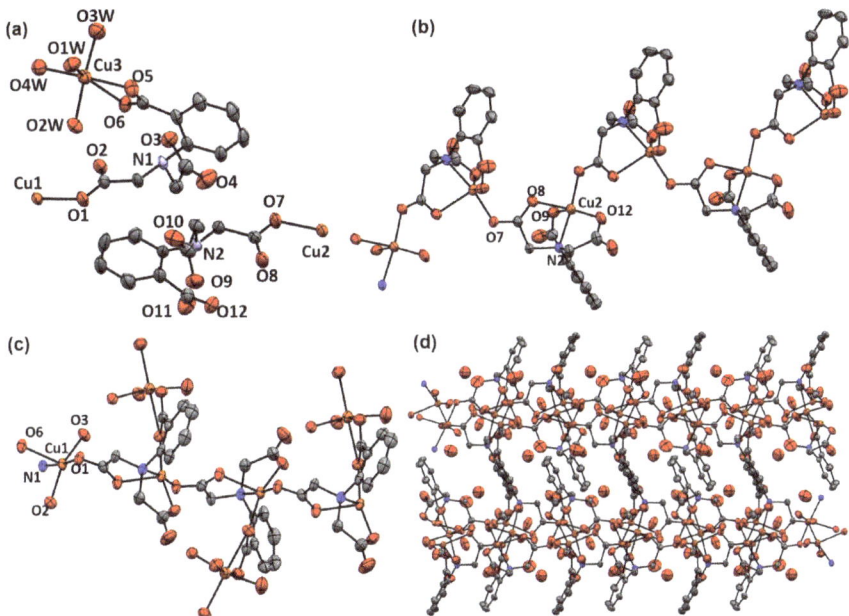

Figure 2. (a) Asymmetric unit of the structure of compound **1** with labeling; (b) chain-like structure of the [Cu(cpida)]$^-$ anion; (c) chain-like structure of the [Cu$_2$(cpida)(H$_2$O)$_4$]$^+$ cation; (d) 1D polymeric structure of compound 1 showing the arrangement of the adjacent chains. Thermal ellipsoids are shown at 50% probability. Hydrogen atoms are omitted for clarity. Color code: O (red), N (blue), carbon (grey), Cu (orange).

The [Cu$_2$(cpida)(H$_2$O)$_4$]$^+$ cation is dinuclear and it is formed by [Cu(cpida)] units connected by carboxylate oxygen atoms. The coordination around Cu1 is also a distorted square pyramid (τ factor = 0.278) [59], with the donor atoms being three oxygen atoms, one of each carboxylate of the cpida^{3-} ligand (O2, O3, and O6), and the nitrogen atom N1 (Figure 2c). The fifth coordination position is occupied by the O1 atom of a neighboring acetic group that is acting as a bridging bidentate group, in a *syn-anti* conformation and with a Cu1–Cu1 separation of 5.17(7) Å. In this way, a 1D chain of [Cu(cpida)] is formed, which is similar to the anionic chain in the structure. The cation presents a second Cu atom (Cu3), which is pentacoordinated, again with a very slightly distorted square pyramid (τ factor = 0.002) [59]. The donor atoms are the oxygen atoms of four water molecules (O1W, O2W, O3W, and O4W) and the O5 atom of a carboxylate group. This carboxylate group, which is directly attached to the phenyl ring, is bridging Cu1 and Cu3 in a *syn-anti* conformation. The equatorial positions of the square pyramid are occupied by the O5 atom and three of the coordinated water molecules (O2W, O3W, and O4W), while the fourth water molecule (O1W) is placed at the axial position. The crystal packing is reinforced by hydrogen bonds between the lattice water molecules and the carboxylate oxygen atoms with an average donor acceptor distance of d(D··A) = 2.74(8) Å and an average D-H···A angle of 161.8° (Table S3, Supporting Information).

Compound **2** was obtained under similar conditions as those used for compound **1**, save for the copper source and diffusion vapors. Selected bond lengths and angles for compound **2** are showed in Table 2 and in Table S1 (Supporting Information). The compound crystallizes in the monoclinic P 2$_1$/n space group (Table 1) and its structure is similar to that of compound **1**, with the difference that just one lattice water molecule and half a molecule of ethanol are present per asymmetric unit (Figure 3a). Compound **2** is also composed of the [Cu$_2$(cpida)(H$_2$O)$_4$]$^+$ cation and the [Cu(cpida)]$^-$ anion, both presenting a chain-like structure. In the [Cu$_2$(cpida)(H$_2$O)$_4$]$^+$ cation, the coordination polyhedron around Cu1 is also a distorted square pyramid (τ factor = 0.281) [59], with the donor atoms being three oxygen atoms of each carboxylate group of the cpida^{3-} anion (O2, O3, and O6, Figure 3a) and the nitrogen atom N1. The fifth coordination position is occupied by the O1 atom of a neighboring acetic group, which is acting as a bridging bidentate group in a *syn-anti* conformation, and with a Cu1–Cu1 separation of 5.17(7) Å. The Cu atom (Cu3) is pentacoordinated, and again with a slightly distorted square pyramid geometry (τ factor = 0.01) [59]. The donor atoms are the oxygen atoms of four water molecules (O1W, O2W, O3W, and O4W) and O5 of the carboxylate group attached to the phenyl ring. This carboxylate group is bridging Cu1 and Cu3 in a *syn-anti* conformation. Three water molecules (O1W, O2W, and O4W) and the O5 atom are occupying the equatorial positions of the square pyramid, while the fourth coordinated water molecule (O3W) is placed at the axial position.

In the [Cu(cpida)]$^-$ anion, the coordination geometry around the Cu2 atom can be described as a slightly distorted square pyramid (τ factor = 0.36) [59], with the donor atoms being the nitrogen atom N2 and three oxygen atoms, one of each carboxylate group of the cpida^{3-} anion (O8, O9, and O12, Figure 3a). The coordination polyhedron is completed by the O-carboxylate of a neighboring [Cu(cpida)] unit. One of the acetic carboxylate groups is acting as a bridging bidentate, in a *syn-anti* conformation and with a Cu–Cu separation of 5.25(7) Å.

The crystal packing is reinforced by hydrogen bonds between the lattice water molecules, the ethanol molecules, and the carboxylate oxygen atoms, with an average donor acceptor distance of d(D···A) = 2.73(5) Å and an average D-H···A angle of 162.2° (Table S4, Supporting Information). An unusual C(4)-H(4B)···O(13)$^{\#1}$ hydrogen bond between the ethanol O(13) atom and the CH$_2$ group of the cationic chain is also present (Table S4).

Given that the unit cells of **1** and **2** showed practically the same geometrical parameters and the same symmetry, we performed an analysis to further characterize these similarities. Using the COMPSTRU feature available on the Bilbao Crystallographic Server (https://www.cryst.ehu.es/ accessed on 14 April 2023), we checked the relationship between both structures, after removing the corresponding solvent molecules. It was found that there is no pair of atoms transformable with a one-angstrom tolerance. Mercury's structure overlay tool (Mercury version 2022.3.0) [63] was also employed, with us observing that when increasing the number of pairs of atoms to overlap, the RMSD between the chosen molecules increased more and more.

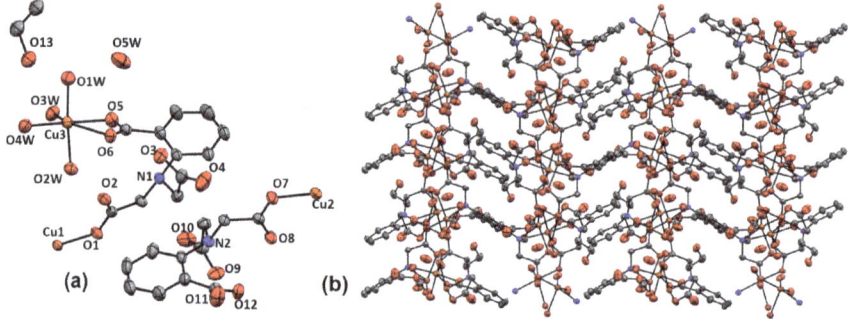

Figure 3. (**a**) Asymmetric unit of the structure of compound **2** with labeling. Thermal ellipsoids are shown at 50% probability. Hydrogen atoms are omitted for clarity. (**b**) Packing of compound **2**, with view along the *c* axis; hydrogen atoms and solvent crystallization molecules are omitted for clarity.

Table 2. Selected bond distances (Å) for the coordination center in compounds **1** and **2** [†].

	Compound 1		2
		Distances	
Cu(1)–O(1)	1.918(3)	Cu(1)–O(1)	1.917(2)
Cu(1)–O(6)[#1]	1.936(3)	Cu(1)–O(6)[#3]	1.933(2)
Cu(1)–O(2)[#1]	2.239(3)	Cu(1)–O(2)[#3]	2.247(2)
Cu(1)–O(3)[#1]	1.943(3)	Cu(1)–O(3)[#3]	1.935(3)
Cu(1)–N(1)[#1]	2.043(3)	Cu(1)–N(1)[#3]	2.043(3)
Cu(2)–O(7)	1.911(3)	Cu(2)–O(8)[#4]	2.238(2)
Cu(2)–O(8)[#2]	2.251(3)	Cu(2)–O(7)	1.905(2)
Cu(2)–O(9)[#2]	1.975(3)	Cu(2)–O(12)[#4]	1.976(3)
Cu(2)–O(12)[#2]	1.959(3)	Cu(2)–O(9)[#4]	1.983(3)
Cu(2)–N(2)[#2]	2.028(3)	Cu(2)–N(2)[#4]	2.036(3)
Cu(3)–O(2W)	1.975(3)	Cu(3)–O(2W)	1.970(3)
Cu(3)–O(5)	1.976(3)	Cu(3)–O(5)	1.990(3)
Cu(3)–O(1W)	2.179(3)	Cu(3)–O(4W)	1.994(3)
Cu(3)–O(4W)	1.984(3)	Cu(3)–O(3W)	2.179(3)
Cu(3)–O(3W)	1.951(3)	Cu(3)–O(1W)	1.946(3)

[†] Symmetry codes: [#1] −x + 1/2, y + 1/2, −z + 3/2; [#2] −x + 1/2, y − 1/2, −z + 5/2; [#3] −x + 3/2, y + 1/2, −z + 3/2; [#4] −x + 1/2, y − 1/2, −z + 3/2.

3.3. Crystal structure of Na[Cu(cpida)] (3)

Compound **3** crystallizes in an orthorhombic *Pbca* space group (Table 1). Selected bond lengths and angles for **3** can be found in Table 3 and Table S2 (Supporting Information). The asymmetric unit (Figure 4a) is composed of the [Cu(cpida)]$^-$ anion with the cpida^{3-} ligand coordinated to the Cu cation in a tridentate manner. The coordination sphere of the copper center is square planar, with the donor atoms being N1, O5, O3, and the O1[#5] atom of a neighboring [Cu(cpida)] unit, with an average Cu-O distance of 1.92(4) Å and a Cu-N distance of 2.046(5) Å. The O1 atom is located in an axial position, at a longer Cu1-O1 distance. Therefore, the [Cu(cpida)] fragments form a chain-like structure, in a zig-zag motif (Figure 4b). In the structure, there are also sodium cations, which are connected to the carboxylate oxygen atoms O5 and O1 and to the O(3)[#6], O(3)[#7], O(6)[#8], O(4)[#6], and O(2)[#5] atoms, connecting the [Cu(cpida)] chains (Figure 4c). The packing of **3** viewed along the b axis is depicted in Figure 4d.

Table 3. Selected bond distances (Å) for the coordination center in compounds **3** and **4** [†].

	Compound 3		4
		Distances	
Cu(1)–O(1)[#5]	1.935(4)	Cu(1)–O(5)	1.903(2)
Cu(1)–O(5)	1.892(4)	Cu(1)–O(3)	1.947(2)
Cu(1)–O(3)	1.940(4)	Cu(1)–O(1)[#9]	1.952(2)
Cu(1)–N(1)	2.046(5)	Cu(1)–O(1)	2.390(3)
Na(1)–O(1)	2.706(5)	Cu(1)–N(1)	2.054(3)
Na(1)–O(5)	2.365(4)	Na(1)–O(5)	2.327(3)
Na(1)–O(3)[#6]	2.903(4)	Na(1)–O(3)[#10]	2.919(3)
Na(1)–O(3)[#7]	2.408(4)	Na(1)–O(3)[#11]	2.371(3)
Na(1)–O(6)[#8]	2.361(5)	Na(1)–O(2)[#9]	2.809(3)
Na(1)–O(4)[#6]	2.397(5)	Na(1)–O(6)[#12]	2.362(3)
Na(1)–O(2)[#5]	2.690(5)	Na(1)–O(4)[#10]	2.401(3)
		Na(1)–O(1)	2.701(3)

[†] Symmetry codes: [#5] −x + 1, y + 1/2, −z + 1/2; [#6] x + 1/2, y, −z + 1/2; [#7] −x + 1, y − 1/2, −z + 1/2; [#8] −x + 3/2, y − 1/2, z; [#9] −x + 1/2, −y + 1/2, z − 1/2; [#10] −x + 1/2, y + 1/2, z; [#11] −x + 1/2, −y + 1/2, z + 1/2; [#12] x, −y + 1, z + 1/2.

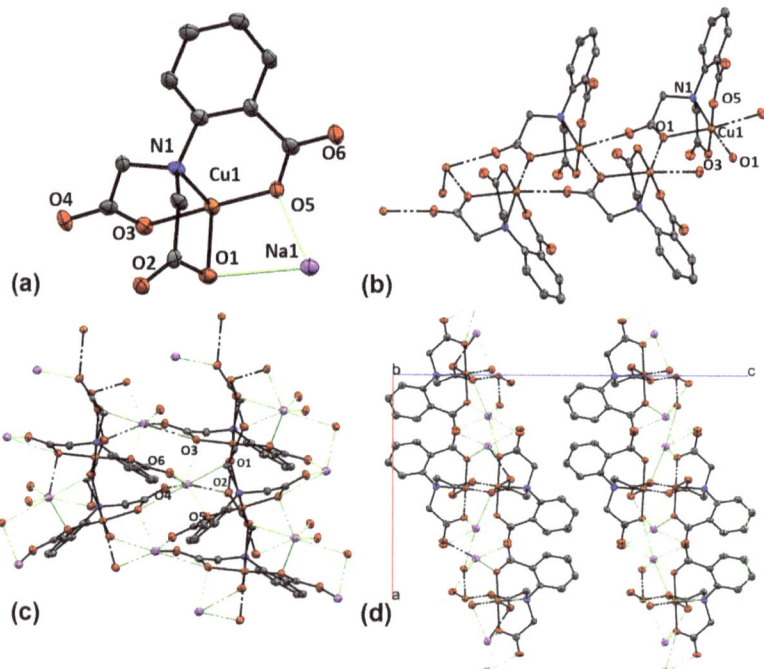

Figure 4. (**a**) Asymmetric unit of compound **3** with labeling. Thermal ellipsoids are shown at 50% probability; (**b**) Zig-zag motif of [Cu(cpida)]⁻ fragments in **3**. Hydrogen atoms and sodium ions are omitted for clarity; (**c**) coordination environment of sodium ions; (**d**) packing of **3** viewed along the b axis. Hydrogen atoms are omitted for clarity.

*3.4. Crystal Structure of Na[Cu(cpida)]·2H$_2$O (**4**)*

Compound **4** crystallizes in the orthorhombic Iba2 space group (Table 1). Selected bond lengths and angles for **4** are displayed in Table 3 and Table S2 (Supporting Information). The composition of the asymmetric unit (Figure 5a) is like that of **3**. It contains the [Cu(cpida)]⁻ anion with the cpida³⁻ ligand coordinated to the Cu cation in a tridentate fashion. The coordination sphere of the copper center is square planar, with the donor atoms being N1, O5, O3, and the O1[#9] atom of a neighboring [Cu(cpida)] unit, with a Cu-N distance of 2.054(3) Å and an average Cu-O distance of 1.93(2) Å. The O1 atom is located in an axial position, at a longer Cu1-O1 distance. Again, the [Cu(cpida)] fragments form a chain-like structure, in a zig-zag motif (Figure 5b). In the structure, there are also sodium cations that are connected to the carboxylate oxygen atoms O5, O1, O(3)[#10], O(3)[#11], O(6)[#12], O(4)[#10], and O(2)[#9], resulting in the overall stabilization of the [Cu(cpida)] chains (Figure 5c). Water crystallization molecules are present in the unit cell but, unfortunately, it was impossible to model it correctly and, therefore, a squeeze method was applied in this case. The analysis of the solvent mask we calculated gave 136 e⁻ in a volume of 446 Å³ in a void per unit cell. This is consistent with the presence of two water molecules per asymmetric unit. The total amount of solvent calculated through this procedure (solvent mask, squeeze) also reasonably matches the elemental analysis and the 8.0% mass loss observed in TGA up to 120 °C (Figure S6), with a calculated mass of 9.6% (two H$_2$O molecules per Na[Cu(cpida)]·2H$_2$O). The packing of **4** is depicted in Figure 5d.

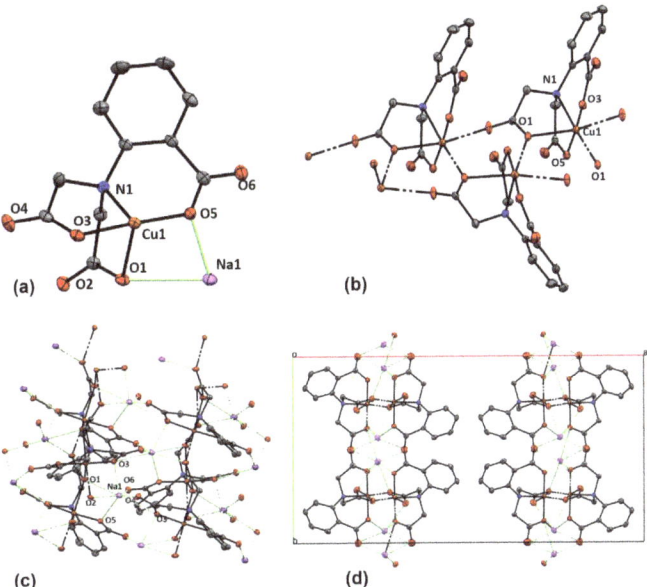

Figure 5. (**a**) Asymmetric unit of the structure of compound **4** with labeling. Thermal ellipsoids are shown at 50% probability. (**b**) Zig-zag motif of [Cu(cpida)]⁻ fragments in **4**. Hydrogen atoms and sodium ions are omitted for clarity. (**c**) Coordination environment of sodium ions. (**d**) Packing of compound **4**, with view along the *c* axis. Hydrogen atoms are omitted for clarity.

4. Conclusions

In this work, we have reported the structure of four new Cu(II) polymeric complexes derived from the 2-(carboxyphenyl)iminodiacetic acid, H_3cpida, obtained under different crystal growing conditions.

Compounds [Cu$_2$(cpida)(H$_2$O)$_4$][Cu(cpida)]·3H$_2$O (**1**) and [Cu$_2$(cpida)(H$_2$O)$_4$][Cu(cpida)]·H$_2$O·1/2EtOH (**2**), obtained through the evaporation of the solvent and ethanol vapor diffusion, respectively, are ionic, formed by the [Cu(cpida)]⁻ anion and the [Cu$_2$(cpida)(H$_2$O)$_4$]⁺ cation, both of them 1D coordination polymers. The structure of the anion is based on [Cu(cpida)] units connected by an acetic oxygen atom that is bridging the Cu centers in a chain-like structure. The [Cu$_2$(cpida)(H$_2$O)$_4$]⁺ cation is dinuclear and it is formed by [Cu(cpida)] units connected by one of the acetic carboxylate groups, while the phenyl carboxylate group is bridging a second copper (II) ion whose coordination sphere is completed with water molecules. Solvent molecules are present in the unit cells of both compounds.

Through diffusion of acetone or methanol vapors, compounds **3**, Na[Cu(cpida)], and **4**, Na[Cu(cpida)]·2H$_2$O, are crystallized. They are also polymeric, with one of the acetic carboxylate groups of the ligand bridging the Cu(II) ions, forming an anionic chain-like structure. The sodium cations are balancing the negative charge, interacting with the oxygen atoms of different carboxylate groups, and connecting the [Cu(cpida)]$_n$ chains.

Supplementary Materials: The following supporting information can be downloaded at: https://www.mdpi.com/article/10.3390/cryst13121669/s1, Figure S1: FT-IR spectrum of compound **1**; Figure S2: FT-IR spectrum of compound **2**; Figure S3: FT-IR spectrum of compound **3**; Figure S4: FT-IR spectrum of compound **4**; Figure S5: TGA diagram of compound **1**; Figure S6: TGA diagram of compound **4**; Table S1. Selected bond angles (°) for the coordination center in compounds **1** and **2**; Table S2. Selected bond angles (°) for the coordination center in compounds **3** and **4**. Table S3: Hydrogen bonds of **1**; Table S4: Hydrogen bonds for **2**.

Author Contributions: Writing—original draft preparation, C.M; writing—review and editing, C.M., S.M., J.G.-P., and C.K.; synthesis and characterization, S.M. and C.M.; diffraction studies, J.G.-P.; conceptualization, C.K and C.M. All authors have read and agreed to the published version of the manuscript.

Funding: This work was partially supported by ANII (Agencia Nacional de Investigación e Innovación, Uruguay) via Project FCE_2_2011_1_N° 6638 and CSIC (Comisión Sectorial de Investigación Científica, Uruguay) through Programa de Apoyo a Grupos de Investigación.

Data Availability Statement: Data are contained within the article and supplementary materials.

Acknowledgments: We are grateful for the support of PEDECIBA (Programa para el Desarrollo de las Ciencias Básicas, Uruguay). J.G.-P. is thankful for the support of MCIN/AEI/10.13039/5011000011033 through the project PID2019-106383GB-C44. C.M. wants to thank Natalia Alvarez (Área Química Inorgánica, DEC, Facultad de Química, Udelar), for the valuable discussion of the results.

Conflicts of Interest: The authors declare no conflict of interest. The funders had no role in the design of the study; in the collection, analyses, or interpretation of data; in the writing of the manuscript; or in the decision to publish the results.

References

1. Ma, L.; Abney, C.; Lin, W. Enantioselective catalysis with homochiral metal–organic frameworks. *Chem. Soc. Rev.* **2009**, *38*, 1248–1256. [CrossRef] [PubMed]
2. Chen, W.; Wu, C. Synthesis, functionalization, and applications of metal–organic frameworks in biomedicine. *Dalton Trans.* **2018**, *47*, 2114–2133. [CrossRef] [PubMed]
3. Baruah, J.B. Coordination polymers in adsorptive remediation of environmental contaminants. *Coord. Chem. Rev.* **2022**, *470*, 214694. [CrossRef]
4. Nguyen, N.T.T.; Nguyen, T.T.T.; Nguyen, D.T.C.; Van Tran, T. Functionalization strategies of metal-organic frameworks for biomedical applications and treatment of emerging pollutants: A review. *Sci. Total Environ.* **2024**, *906*, 167295. [CrossRef]
5. Dutta, M.; Bora, J.; Chetia, B. Overview on recent advances of magnetic metal–organic framework (MMOF) composites in removal of heavy metals from aqueous system. *Environ. Sci. Pollut. Res.* **2022**, *30*, 13867–13908. [CrossRef]
6. Li, J.-R.; Kuppler, R.J.; Zhou, H.-C. Selective gas adsorption and separation in metal–organic frameworks. *Chem. Soc. Rev.* **2009**, *38*, 1477–1504. [CrossRef]
7. Farrusseng, D.; Aguado, S.; Pinel, C. Metal–Organic Frameworks: Opportunities for Catalysis. *Angew. Chem. Int. Ed.* **2009**, *48*, 7502–7513. [CrossRef]
8. Ma, Z.; Moulton, B. Recent advances of discrete coordination complexes and coordination polymers in drug delivery. *Coord. Chem. Rev.* **2011**, *255*, 1623–1641. [CrossRef]
9. Kurmoo, M. Magnetic metal–organic frameworks. *Chem. Soc. Rev.* **2009**, *38*, 1353–1379. [CrossRef]
10. Jiang, H.-L.; Xu, Q. Porous metal–organic frameworks as platforms for functional applications. *Chem. Commun.* **2011**, *47*, 3351–3370. [CrossRef]
11. Murray, L.J.; Dincă, M.; Long, J.R. Hydrogen storage in metal–organic frameworks. *Chem. Soc. Rev.* **2009**, *38*, 1294–1314. [CrossRef] [PubMed]
12. Loukopoulos, E.; Kostakis, G.E. Review: Recent advances of one-dimensional coordination polymers as catalysts. *J. Coord. Chem.* **2018**, *71*, 371–410. [CrossRef]
13. Pettinari, C.; Marchetti, F.; Mosca, N.; Tosi, G.; Drozdov, A. Application of metal–Organic frameworks. *Polym. Int.* **2017**, *66*, 731–744. [CrossRef]
14. Batten, S.R.; Champness, N.R.; Chen, X.-M.; Garcia-Martinez, J.; Kitagawa, S.; Öhrström, L.; O'keeffe, M.; Suh, M.P.; Reedijk, J. Terminology of metal–organic frameworks and coordination polymers (IUPAC Recommendations 2013). *Pure Appl. Chem.* **2013**, *85*, 1715–1724. [CrossRef]
15. Batten, S.R.; Champness, N.R.; Chen, X.-M.; Garcia-Martinez, J.; Kitagawa, S.; Öhrström, L.; O'Keeffe, M.; Suh, M.P.; Reedijk, J. Coordination polymers, metal–organic frameworks and the need for terminology guidelines. *CrystEngComm* **2012**, *14*, 3001–3004. [CrossRef]
16. Yusuf, V.F.; Malek, N.I.; Kailasa, S.K. Review on Metal–Organic Framework Classification, Synthetic Approaches, and Influencing Factors: Applications in Energy, Drug Delivery, and Wastewater Treatment. *ACS Omega* **2022**, *7*, 44507–44531. [CrossRef]
17. Masoomi, M.Y.; Morsali, A. Applications of metal–organic coordination polymers as precursors for preparation of nano-materials. *Coord. Chem. Rev.* **2012**, *256*, 2921–2943. [CrossRef]
18. Robson, R. Design and its limitations in the construction of bi- and poly-nuclear coordination complexes and coordination polymers (aka MOFs): A personal view. *Dalton Trans.* **2008**, *38*, 5113–5131. [CrossRef]
19. Zhang, M.-L.; Zheng, Y.-J.; Ma, Z.-Z.; Ren, Y.-X.; Cao, J.; Wang, Z.-X.; Wang, J.-J. Zinc(II) and cadmium(II) complexes of long flexible bis(imidazole) and phenylenediacetate ligands, synthesis, structure, and luminescent property. *Polyhedron* **2018**, *146*, 180–186. [CrossRef]

20. Khalaj, M.; Lalegani, A.; Akbari, J.; Ghazanfarpour-Darjani, M.; Lyczko, K.; Lipkowski, J. Synthesis and characterization of three new Cd(II) coordination polymers with bidentate flexible ligands: Formation of 3D and 1D structures. *J. Mol. Struct.* **2018**, *1169*, 31–38. [CrossRef]
21. Chen, M.-L.; Zhou, Z.-H. Structural diversity of 1,3-propylenediaminetetraacetato metal complexes: From coordination monomers to coordination polymers and MOF materials. *Inorg. Chim. Acta* **2017**, *458*, 199–217. [CrossRef]
22. Li, H.; Tian, H.; Guo, M.; He, F.Y.; Hu, C. Self-complementary self-assembly of 3D coordination framework involving d-block, s-block metal ions and flexible NTA^{3-} ligand. *Inorg. Chem. Commun.* **2006**, *9*, 895–898. [CrossRef]
23. Liu, T.; Lü, J.; Shi, L.; Guo, Z.; Cao, R. Conformation control of a flexible 1,4-phenylenediacetate ligand in coordination complexes: A rigidity-modulated strategy. *CrystEngComm* **2008**, *11*, 583–588. [CrossRef]
24. Liu, T.-F.; Lü, J.; Cao, R. Coordination polymers based on flexible ditopic carboxylate or nitrogen-donor ligands. *CrystEngComm* **2010**, *12*, 660–670. [CrossRef]
25. Mateescu, A.; Gabriel, C.; Raptis, R.G.; Baran, P.; Salifoglou, A. pH—Specific synthesis, spectroscopic, and structural characterization of an assembly of species between Co(II) and N,N-bis(phosphonomethyl)glycine. Gaining insight into metal-ion phosphonate interactions in aqueous Co(II)–organophosphonate systems. *Inorg. Chim. Acta* **2007**, *360*, 638–648. [CrossRef]
26. Zhang, Q.-Z.; Lu, C.-Z.; Yang, W.-B.; Yu, Y.-Q. A novel three-dimensional framework formed by polymeric aqua (nitrilotriacetato)-lanthanum: [La($C_6H_6NO_6$)H_2O]$_n$. *Inorg. Chem. Commun.* **2004**, *7*, 277–279. [CrossRef]
27. Carballo, R.; Covelo, B.; El Fallah, M.S.; Ribas, J.; Vázquez-López, E.M. Supramolecular Architectures and Magnetic Behavior of Coordination Polymers from Copper(II) Carboxylates and 1,2-Bis(4-pyridyl)ethane as a Flexible Bridging Ligand. *Cryst. Growth Des.* **2007**, *7*, 1069–1077. [CrossRef]
28. Hawes, C.S.; Chilton, N.F.; Moubaraki, B.; Knowles, G.P.; Chaffee, A.L.; Murray, K.S.; Batten, S.R.; Turner, D.R. Coordination polymers from a highly flexible alkyldiamine-derived ligand: Structure, magnetism and gas adsorption studies. *Dalton Trans.* **2015**, *44*, 17494–17507. [CrossRef]
29. Lai, L.-Y.; Liu, Z.; Han, G.-C.; Chen, Z. Synthesis, Crystal Structure and Properties of Three Metal Complexes Based on a Flexible Schiff Base Ligand. *J. Clust. Sci.* **2015**, *26*, 1845–1855. [CrossRef]
30. Li, H.-Y.; Cao, L.-H.; Wei, Y.-L.; Xu, H.; Zang, S.-Q. Construction of a series of metal–organic frameworks based on novel flexible ligand 4-carboxy-1-(3,5-dicarboxy-benzyl)-pyridinium chloride and selective d-block metal ions: Crystal structures and photoluminescence. *CrystEngComm* **2015**, *17*, 6297–6307. [CrossRef]
31. Karmakar, A.; Oliver, C.L.; Roy, S.; Öhrström, L. The synthesis, structure, topology and catalytic application of a novel cubane-based copper(ii) metal–organic framework derived from a flexible amido tripodal acid. *Dalton Trans.* **2015**, *44*, 10156–10165. [CrossRef]
32. Yu, M.; Xie, L.; Liu, S.; Wang, C.; Cheng, H.; Ren, Y.; Su, Z. Photoluminescent metal-organic framework with hex topology constructed from infinite rod-shaped secondary building units and single e,e-trans-1,4-cyclohexanedicarboxylic dianion. *Inorg. Chim. Acta* **2007**, *360*, 3108–3112. [CrossRef]
33. Xie, Y.; Bai, F.; Xing, Y.; Wang, Z.; Zhao, H.; Shi, Z. Synthesis and Crystal Structure of Novel Coordination Polymers with Nitrilotripropionic Acid. *Z. Anorg. Allg. Chem.* **2010**, *636*, 1585–1590. [CrossRef]
34. Huang, R.-W.; Li, B.; Zhang, Y.-Q.; Zhao, Y.; Zang, S.-Q.; Xu, H. Divalent cobalt, zinc, and copper coordination polymers based on a new bifunctional ligand: Syntheses, crystal structures, and properties. *Inorg. Chem. Commun.* **2014**, *39*, 106–109. [CrossRef]
35. Chai, X.; Zhang, H.; Zhang, S.; Cao, Y.; Chen, Y. The tunable coordination architectures of a flexible multicarboxylate N-(4-carboxyphenyl)iminodiacetic acid via different metal ions, pH values and auxiliary ligand. *J. Solid State Chem.* **2009**, *182*, 1889–1898. [CrossRef]
36. Pan, Z.; Zheng, H.; Wang, T.; Song, Y.; Li, Y.; Guo, Z.; Batten, S.R. Hydrothermal Synthesis, Structures, and Physical Properties of Four New Flexible Multicarboxylate Ligands-Based Compounds. *Inorg. Chem.* **2008**, *47*, 9528–9536. [CrossRef]
37. Braña, E.; Mendoza, C.; Vitoria, P.; González-Platas, J.; Domínguez, S.; Kremer, C. Trinuclear Cu(II) capsules self-assembled by nitrilotripropionate. *Inorg. Chim. Acta* **2014**, *417*, 192–200. [CrossRef]
38. Martínez, S.; Veiga, N.; Torres, J.; Kremer, C.; Mendoza, C. Polynuclear complexes in solution: An experimental and theoretical study on the interaction of nitrilotripropionate anion with metal ions. *Inorg. Chim. Acta* **2018**, *483*, 53–60. [CrossRef]
39. Braña, E.; Quiñone, D.; Martínez, S.; Grassi, J.; Carrera, I.; Torres, J.; González-Platas, J.; Seoane, G.; Kremer, C.; Mendoza, C. New polynuclear compounds based on N-benzyliminodipropionic acid: Solution studies, synthesis, and X-ray crystal structures. *J. Coord. Chem.* **2016**, *69*, 3650–3663. [CrossRef]
40. Bera, M.; Musie, G.T.; Powell, D.R. Zinc(II) mediated cyclization and complexation of an unsymmetrical dicarboxyamine ligand: Synthesis, spectral and crystal structures characterizations. *Inorg. Chem. Commun.* **2008**, *11*, 293–299. [CrossRef]
41. Palanisami, N.S.K.K.; Gopalakrishnan, T.; Moon, I.-S. A mixed Ni(II) ionic complex containing V-shaped water trimer: Synthesis, spectral, structural and thermal properties of {[Ni(2,2'-bpy)$_3$][Ni(2-cpida)(2,2'-bpy)](ClO$_4$).3H$_2$O. *J. Chem. Sci.* **2015**, *127*, 873–876. [CrossRef]
42. Tomita, T.; Kyuno, E.; Tsuchiya, R. The Chromium(III) Complexes with Anthranilicdiacetic Acid. *Bull. Chem. Soc. Jpn.* **1969**, *42*, 947–951. [CrossRef]
43. Chatterjee, C.; Singh, R.S.; Phulambrikar, A.; Das, S. Synthesis and spectral characterization of cobalt(III) complexes of N-(o-carboxyphenyl)iminodiacetic acid. *J. Chem. Soc. Dalton Trans.* **1988**, *1988*, 159–162. [CrossRef]

44. Ma, J.; Jiang, F.; Zhou, K.; Chen, L.; Wu, M.; Hong, M. Effects of Temperature and Anion on the Copper(II) Complexes based on 2-(Carboxyphenyl)iminodiacetic Acid and 1,10-Phenanthroline. *Z. Anorg. Und Allg. Chem.* **2015**, *641*, 1998–2004. [CrossRef]
45. Nallasamy, P.; Senthilkumar, K.; Mohan, G.; Moon, I.-S. Structural and luminescent properties of a tetranuclear cage-type cadmium(II) carboxylate cluster containing a V-shaped water trimer. *J. Coord. Chem.* **2016**, *69*, 1005–1013. [CrossRef]
46. Murugesu, M.; Anson, C.E.; Powell, A.K. Engineering of ferrimagnetic Cu_{12}-cluster arrays through supramolecular interactions. *Chem. Commun.* **2002**, *10*, 1054–1055. [CrossRef] [PubMed]
47. Murugesu, M.; King, P.; Clérac, R.; Anson, C.E.; Powell, A.K. A novel nonanuclear Cu^{II} carboxylate-bridged cluster aggregate with an S = 7/2 ground spin state. *Chem. Commun.* **2004**, *6*, 740–741. [CrossRef]
48. Murugesu, M.; Clérac, R.; Anson, C.E.; Powell, A.K. Polycopper(II) aggregates as building blocks for supramolecular magnetic structures. *J. Phys. Chem. Solids* **2004**, *65*, 667–676. [CrossRef]
49. McCowan, C.S.; Groy, T.L.; Caudle, M.T. Synthesis, Structure, and Preparative Transamination of Tetrazinc Carbamato Complexes Having the Basic Zinc Carboxylate Structure. *Inorg. Chem.* **2002**, *41*, 1120–1127. [CrossRef]
50. Bandyopadhyay, S.; Das, A.; Mukherjee, G.; Cantoni, A.; Bocelli, G.; Chaudhuri, S.; Ribas, J. Synthesis, X-ray crystal structures, physicochemical and magnetic studies on copper(II) N-(2-carboxyphenyl) iminodiacetate polymers. *Polyhedron* **2004**, *23*, 1081–1088. [CrossRef]
51. Yong, G.-P.; Qiao, S.; Wang, Z.-Y.; Cui, Y. Five-, seven-, and eight-coordinate Cd(II) coordination polymers built by anthranilic acid derivatives: Synthesis, structures and photoluminescence. *Inorg. Chim. Acta* **2005**, *358*, 3905–3913. [CrossRef]
52. Chu, Q.; Liu, G.-X.; Okamura, T.-A.; Huang, Y.-Q.; Sun, W.-Y.; Ueyama, N. Structure modulation of metal–organic frameworks via reaction pH: Self-assembly of a new carboxylate containing ligand N-(3-carboxyphenyl)iminodiacetic acid with cadmium(II) and cobalt(II) salts. *Polyhedron* **2008**, *27*, 812–820. [CrossRef]
53. Yong, G.; Wang, Z.; Cui, Y. Synthesis, structural characterization and properties of copper(II) and zinc(II) coordination polymers with a new bridging chelating ligand. *Eur. J. Inorg. Chem.* **2004**, *2004*, 4317–4323. [CrossRef]
54. Yong, G.; Wang, Z.; Chen, J. Two-dimensional and three-dimensional nickel(II) supramolecular complexes based on the new chelating ligand N-(4-carboxyphenyl)iminodiacetic acid: Hydrothermal synthesis and crystal structures. *J. Mol. Struct.* **2004**, *707*, 223–229. [CrossRef]
55. *CrysAlisPro Software System*, 1.171.39.46; Rigaku Corporation: Oxford, UK, 2018.
56. Sheldrick, G.M. *SHELXT*—Integrated space-group and crystal-structure determination. *Acta Crystallogr. Sect. A Found. Adv.* **2015**, *71*, 3–8. [CrossRef] [PubMed]
57. Sheldrick, G.M. Crystal structure refinement with SHELXL. *Acta Crystallogr. Sect. C Struct. Chem.* **2015**, *71*, 3–8. [CrossRef] [PubMed]
58. Dolomanov, O.V.; Bourhis, L.J.; Gildea, R.J.; Howard, J.A.K.; Puschmann, H. OLEX2: A complete structure solution, refinement and analysis program. *J. Appl. Crystallogr.* **2009**, *42*, 339–341. [CrossRef]
59. Addison, A.W.; Rao, T.N.; Reedijk, J.; van Rijn, J.; Verschoor, G.C. Synthesis, structure, and spectroscopic properties of copper(II) compounds containing nitrogen–sulphur donor ligands; the crystal and molecular structure of aqua [1,7-bis(N-methylbenzimidazol-2′-yl)-2,6-dithiaheptane]copper(II) perchlorate. *J. Chem. Soc. Dalton Trans.* **1984**, 1349–1356. [CrossRef]
60. Choquesillo-Lazarte, D.; Covelo, B.; González-Pérez, J.M.; Castiñeiras, A.; Niclós-Gutiérrez, J. Metal chelates of N-(2-pyridylmethyl)iminodiacetate(2-) ion (pmda). Part I. Two mixed-ligand copper(II) complexes of pmda with N,N-chelating bases. Synthesis, crystal structure and properties of $H_2pmda·0.5H_2O$, [Cu(pmda)(pca)]·$3H_2O$ (pca=α-picolylamine) and [Cu(pmda)(Hpb)]·$5H_2O$ (Hpb=2-(2′-pyridyl)benzimidazole). *Polyhedron* **2002**, *21*, 1485–1495. [CrossRef]
61. Román-Alpiste, M.; Martín-Ramos, J.; Castiñeiras-Campos, A.; Bugella-Altamirano, E.; Sicilia-Zafra, A.; González-Pérez, J.; Niclós-Gutiérrez, J. Synthesis, XRD structures and properties of diaqua(iminodiacetato)copper(II), [Cu(IDA)($H_2O)_2$], and aqua(benzimidazole)(iminodiacetato)copper(II), [Cu(IDA)(HBzIm)(H_2O)]. *Polyhedron* **1999**, *18*, 3341–3351. [CrossRef]
62. Bugella-Altamirano, E.; González-Pérez, J.M.; Choquesillo-Lazarte, D.; Carballo, R.; Castiñeiras, A.; Niclós-Gutiérrez, J. A structural evidence for the preferential coordination of the primary amide group versus the unionised carboxyl group: Synthesis, molecular and crystal structure, and properties of [Cu(HADA)$_2$], a new copper(II) bis-chelate (H_2ADA=N-(2-carbamoylmethyl)iminodiacetic acid). *Inorg. Chem. Commun.* **2003**, *6*, 71–73. [CrossRef]
63. Macrae, C.F.; Sovago, I.; Cottrell, S.J.; Galek, P.T.A.; McCabe, P.; Pidcock, E.; Platings, M.; Shields, G.P.; Stevens, J.S.; Towler, M.; et al. *Mercury 4.0*: From visualization to analysis, design and prediction. *J. Appl. Crystallogr.* **2020**, *53*, 226–235. [CrossRef] [PubMed]

Disclaimer/Publisher's Note: The statements, opinions and data contained in all publications are solely those of the individual author(s) and contributor(s) and not of MDPI and/or the editor(s). MDPI and/or the editor(s) disclaim responsibility for any injury to people or property resulting from any ideas, methods, instructions or products referred to in the content.

Article

Dinuclear Molybdenum(VI) Complexes Based on Flexible Succinyl and Adipoyl Dihydrazones

Edi Topić [1], Vladimir Damjanović [2], Katarina Pičuljan [1] and Mirta Rubčić [1,*]

[1] Department of Chemistry, Faculty of Science, University of Zagreb, Horvatovac 102a, 10000 Zagreb, Croatia; edi.topic@chem.pmf.hr (E.T.); kpiculjan@chem.pmf.hr (K.P.)
[2] Department of Chemistry and Biochemistry, School of Medicine, University of Zagreb, Šalata 3, 10000 Zagreb, Croatia; vladimir.damjanovic@mef.hr
* Correspondence: mirta@chem.pmf.hr; Tel.: +385-1-4606-374

Abstract: A series of molybdenum(VI) complexes with aryl-functionalized alkyl dihydrazones was prepared by the reaction of [MoO$_2$(acac)$_2$] and the appropriate dihydrazone in methanol. Their solid-state structures were elucidated via single-crystal X-ray diffraction (SC-XRD) and Fourier-transform infra-red (FTIR) spectroscopy, while the thermal stability of compounds was inspected by combined thermogravimetric analysis (TGA) and differential scanning calorimetry (DSC) experiments. The behaviour of complexes in DMSO-d_6 solution was explored by nuclear magnetic resonance (NMR). The relevant data show that all complexes are dinuclear, with dihydrazones acting as ditopic hexadentate ligands. The in vitro cytotoxic activity of the prepared molybdenum(VI) complexes was evaluated on THP-1 and HepG2 cell lines, while their antibacterial activity was tested against *Staphylococcus aureus*, *Enterococcus faecalis*, *Escherichia coli*, and *Moraxella catarrhalis* bacteria. The majority of compounds proved to be non-cytotoxic, while some exhibited superior antibacterial activity in comparison to dihydrazone ligands.

Keywords: hydrazones; Mo(VI) complexes; structural analysis; NMR spectroscopy; cytotoxic and antibacterial activity

Citation: Topić, E.; Damjanović, V.; Pičuljan, K.; Rubčić, M. Dinuclear Molybdenum(VI) Complexes Based on Flexible Succinyl and Adipoyl Dihydrazones. *Crystals* 2024, 14, 135. https://doi.org/10.3390/cryst14020135

Academic Editor: Alexander Y. Nazarenko

Received: 5 January 2024
Revised: 23 January 2024
Accepted: 25 January 2024
Published: 29 January 2024

Copyright: © 2024 by the authors. Licensee MDPI, Basel, Switzerland. This article is an open access article distributed under the terms and conditions of the Creative Commons Attribution (CC BY) license (https://creativecommons.org/licenses/by/4.0/).

1. Introduction

Over the last few decades, hydrazones have emerged as the privileged class of ligands owing to their stability, acid-base properties, and structural modularity [1]. Hydrazones are recognized for their remarkable coordination chemistry, which arises from their ability to act as flexible ligands, forming stable complexes with various metals [2]. This flexibility is attributed to the hydrazone's nitrogen and oxygen atoms, which can engage in both chelating and bridging modes in metal–organic frameworks (MOFs) [3]. Such features make them suitable for the development of metal–organic assemblies for specialized applications related to, e.g., magnetism, catalysis [4], or biomedicine [5–8]. Furthermore, the prospect of E/Z isomerization, particularly as a response to different stimuli, such as light, pH, or heat, renders these systems suitable for the design of molecular switches or even more complex stimuli-responsive metal–organic architectures [9,10]. On the other hand, appropriate modification of the hydrazonic scaffolds can give rise to metal or covalent–organic frameworks for practical applications such as gas storage or separations [11].

By introducing more than one hydrazonic functionality within the same ligand molecule, and thus enhancing its coordination potential, one can target more complex metal-organic structures and functions as compared to the monohydrazone counterparts [12]. Multihydrazones, particularly those with alkyl chains, exhibit enhanced flexibility and can form more intricate and dynamic structures [13]. A notable example is the alkyl dihydrazone-based multinuclear Cu cages, which have been studied for their unique magnetic properties [14,15]. Structures of this type have also shown properties like magnetic refrigeration and slow magnetic relaxation, which could have a number of promising applications [16].

Complexation with appropriate metal cations is often beneficial when one aims to modulate the bioactivity of the related organic entities [17,18]. For monohydrazones, whose antibacterial, antifungal, and antitumor properties have been widely acclaimed [19–22], derivatization via metal cation coordination has proven to be a viable route towards complexes with altered or enhanced biological properties [23]. The flexible nature of succinyl and adipoyl dihydrazones and its influence on the properties of the corresponding complexes, coupled with the fact that dihydazones have been considerably less investigated as bioactive compounds [24], motivated us to explore the cytotoxic and antibacterial activity of dihydrazone-based Mo(VI) complexes and compare it with those of the free ligands [25].

Here, we present a solid-state and a solution study of a series of dinuclear molybdenum(VI) complexes with aryl-functionalized alkyl dihydrazones. Namely, we provide simple synthetic routes towards the title compounds, accompanied by their thorough solid-state analysis via single-crystal X-ray diffraction (SCXRD), Fourier-transform infra-red (FTIR) spectroscopy, and simultaneous thermogravimetry and differential scanning calorimetry (TGA-DSC). In DMSO-d_6 solution, complexes were explored by nuclear magnetic resonance (NMR), unveiling symmetrical structures comparable with those established in the solid state. Finally, we focused on an evaluation of the cytotoxic and antibacterial activities of the obtained compounds against selected human cancer cell lines, and Gram-positive and Gram-negative bacterial strains, respectively.

2. Materials and Methods

2.1. Synthesis

Succinic dihydrazones of salicylaldehyde (**H$_4$L^1**), 2-hydroxy-1-naphtaldehyde(**H$_4$L^2**) 2,3-dihydroxybenzaldehyde (**H$_4$L^3**), and 2,4-dihydroxybenzaldehyde (**H$_4$L^4**), as well as adipic dihydrazones of salicylaldehyde (**H$_4$L^5**), 2-hydroxy-1-naphtaldehyde (**H$_4$L^6**) 2,3-dihydroxybenzaldehyde (**H$_4$L^7**), and 2,4-dihydroxybenzaldehyde (**H$_4$L^8**) were prepared by a previously published procedure [15]. [MoO$_2$(acac)$_2$] was synthesized by a well-established synthetic protocol [26]. Chemicals for the synthesis were purchased from TCI and used as received. Methanol, used in syntheses, was purchased from Kemika (Zagreb, Croatia).

In general, complexes were prepared by the reaction of two equivalents of [MoO$_2$(acac)$_2$] and one equivalent of selected dihydrazone suspended in methanol (Supplementary Materials, Scheme S1). The resulting suspensions were refluxed for two hours, during which the ligands slowly reacted. Upon cooling, the crystalline material that was deposited was filtered, washed with a small amount of cold methanol, and dried in air.

2.1.1. Synthesis of [Mo$_2$O$_4$(MeOH)$_2$(L^1)]·2MeOH

Obtained by the reaction of 0.5 mmol (175 mg) of **H$_4$L^1** and 1.0 mmol (326 mg) of [MoO$_2$(acac)$_2$] in 20 mL of methanol. Orange powder. Yield: 0.32 g (88%). Anal. Calcd. For Mo$_2$C$_{22}$H$_{30}$N$_4$O$_8$ (734.42): C, 35.98%, H, 4.12%, N, 7.63%, found: C, 35.98%, H, 4.12%, N, 7.86%. IR spectroscopy: 1615, 1599 υ(C=N); 1556, 1541 υ(C=C); 1338 υ(C–O$_{en}$); 1270 υ(C–O$_{phen}$); 1011 υ(C–O$_{MeOH}$); 935 υ_{sym}(MoO$_2$); 907 υ_{asym}(MoO$_2$). TGA analysis: MeOH calcd. 17.45%, found 17.84%; MoO$_3$ calcd. 39.20%, found 36.83%.

2.1.2. Synthesis of [Mo$_2$O$_4$(MeOH)$_2$(L^2)]

Obtained by the reaction of 0.5 mmol (225 mg) of **H$_4$L^2** and 1.0 mmol (326 mg) of [MoO$_2$(acac)$_2$] in 20 mL of methanol. Yellow-orange powder. Yield: 0.29 g (76%). Anal. Calcd. For Mo$_2$C$_{28}$H$_{26}$N$_4$O$_6$ (770.45): C, 43.65%, H, 3.4%, N, 7.27%, found: C, 41.47%, H, 3.37%, N, 7.05%. IR spectroscopy: 1615, 1596 υ(C=N); 1550, 1534 υ(C=C); 1329 υ(C–O$_{en}$); 1278 υ(C–O$_{phen}$); 1012 υ(C–O$_{MeOH}$); 934 υ_{sym}(MoO$_2$); 907 υ_{asym}(MoO$_2$). TGA analysis: MeOH calcd. 8.32%, found 9.80%; MoO$_3$ calcd. 37.37%, found 34.26%.

2.1.3. Synthesis of [Mo$_2$O$_4$(MeOH)$_2$(L^3)]·2MeOH

Obtained by the reaction of 0.5 mmol (191 mg) of H$_4$L^3 and 1.0 mmol (326 mg) of [MoO$_2$(acac)$_2$] in 20 mL of methanol. Red powder. Yield: 0.33 g (93%). Anal. Calcd. For Mo$_2$C$_{20}$H$_{22}$N$_4$O$_8$ (702.33): C, 34.2%, H, 3.16%, N, 7.98%, found: C, 33.52%, H, 3.29%, N, 7.98%. IR spectroscopy: 1603 υ(C=N); 1568, 1538 υ(C=C); 1340 υ(C–O$_{en}$); 1260 υ(C–O$_{phen}$); 1015 υ(C–O$_{MeOH}$); 933 υ_{sym}(MoO$_2$); 901 υ_{asym}(MoO$_2$). TGA analysis: MeOH calcd. 16.72%, found 17.09%; MoO$_3$ calcd. 37.56%, found 36.23%.

2.1.4. Synthesis of [Mo$_2$O$_4$(MeOH)$_2$(L^4)]

Obtained by the reaction of 0.5 mmol (191 mg) of H$_4$L^4 and 1.0 mmol (326 mg) of [MoO$_2$(acac)$_2$] in 20 mL of methanol. Red powder. Yield: 0.29 g (76%). Anal. Calcd. For Mo$_2$C$_{22}$H$_{30}$N$_4$O$_{10}$ (766.41): C, 34.48%, H, 3.95%, N, 7.31%, found: C, 35.17%, H, 4.11%, N, 7.31%. IR spectroscopy: 1609 υ(C=N); 1571, 1549 υ(C=C); 1334, 1315 υ(C–O$_{en}$, C–O$_{phen}$); 1014 υ(C–O$_{MeOH}$); 944 υ_{sym}(MoO$_2$); 873 υ_{asym}(MoO$_2$). TGA analysis: MeOH calcd. 9.12%, found 8.54%; MoO$_3$ calcd. 40.99%, found 39.99%.

2.1.5. Synthesis of [Mo$_2$O$_4$(MeOH)$_2$(L^5)]

Obtained by the reaction of 0.5 mmol (189 mg) of H$_4$L^5 and 1.0 mmol (326 mg) of [MoO$_2$(acac)$_2$] in 20 mL of methanol. Orange powder. Yield: 0.28 g (80%). Anal. Calcd. For Mo$_2$C$_{22}$H$_{26}$N$_4$O$_6$ (698.38): C, 37.84%, H, 3.75%, N, 8.02%, found: C, 38.22%, H, 3.6%, N, 7.94%. IR spectroscopy: 1611, 1596 υ(C=N); 1558, 1546 υ(C=C); 1321 υ(C–O$_{en}$); 1284, 1270 υ(C–O$_{phen}$); 1010 υ(C–O$_{MeOH}$); 935 υ_{sym}(MoO$_2$); 880 υ_{asym}(MoO$_2$). TGA analysis: MeOH calcd. 9.18%, found 10.15%; MoO$_3$ calcd. 41.22%, found 39.30%.

2.1.6. Synthesis of [Mo$_2$O$_4$(MeOH)$_2$(L^6)]

Obtained by the reaction of 0.5 mmol (239 mg) of H$_4$L^6 and 1.0 mmol (326 mg) of [MoO$_2$(acac)$_2$] in 20 mL of methanol. Yellow-orange powder. Yield: 0.38 g (95%). Anal. Calcd. For Mo$_2$C$_{30}$H$_{30}$N$_4$O$_6$ (798.5): C, 45.13%, H, 3.79%, N, 7.02%, found: C, 44.68%, H, 3.64%, N, 6.67%. IR spectroscopy: 1618, 1599 υ(C=N); 1551, 1533 υ(C=C); 1329 υ(C–O$_{en}$); 1277 υ(C–O$_{phen}$); 1000 υ(C–O$_{MeOH}$); 937 υ_{sym}(MoO$_2$); 904 υ_{asym}(MoO$_2$). TGA analysis: MeOH calcd. 8.03%, found 9.78%; MoO$_3$ calcd. 36.05%, found 35.89%.

2.1.7. Synthesis of [Mo$_2$O$_4$(MeOH)$_2$(L^7)]

Obtained by the reaction of 0.5 mmol (205 mg) of H$_4$L^7 and 1.0 mmol (326 mg) of [MoO$_2$(acac)$_2$] in 20 mL of methanol. Red powder. Yield: 0.29 g (79%). Anal. Calcd. For Mo$_2$C$_{22}$H$_{26}$N$_4$O$_8$ (730.38): C, 36.18%, H, 3.59%, N, 7.67%, found: C, 35.82%, H, 3.66%, N, 8.05%. IR spectroscopy: 1603 υ(C=N); 1569, 1551 υ(C=C); 1324 υ(C–O$_{en}$); 1263 υ(C–O$_{phen}$); 1035 υ(C–O$_{MeOH}$); 939 υ_{sym}(MoO$_2$); 907, 878 υ_{asym}(MoO$_2$). TGA analysis: MeOH calcd. 8.77%, found 9.11%; MoO$_3$ calcd. 39.42%, found 38.07%.

2.1.8. Synthesis of [Mo$_2$O$_4$(MeOH)$_2$(L^8)]·2MeOH

Obtained by the reaction of 0.5 mmol (205 mg) of H$_4$L^8 and 1.0 mmol (326 mg) of [MoO$_2$(acac)$_2$] in 20 mL of methanol. Red powder. Yield: 0.33 g (84%). Anal. Calcd. For Mo$_2$C$_{24}$H$_{34}$N$_4$O$_{10}$ (794.46): C, 36.28%, H, 4.31%, N, 7.05%, found: C, 37.01%, H, 4.44%, N, 7.33%. IR spectroscopy: 1596 υ(C=N); 1568, 1553 υ(C=C); 1333 υ(C–O$_{en}$); 1292 υ(C–O$_{phen}$); 1016 υ(C–O$_{MeOH}$); 940 υ_{sym}(MoO$_2$); 885 υ_{asym}(MoO$_2$). TGA analysis: MeOH calcd. 16.13%, found 16.13%; MoO$_3$ calcd. 36.24%, found 34.87%.

2.2. Methods

The chemical composition analysis for carbon, hydrogen, and nitrogen was conducted at the Analytical Services Laboratory, Ruđer Bošković Institute (Zagreb, Croatia). X-ray powder diffraction patterns of the samples were obtained using an Empyrean diffractometer (Malvern Panalytical, Almelo, The Netherlands) utilizing copper Kα radiation.

Measurements were made using zero-background holders and the Bragg−Brentano setup, covering a 2θ range of $4°$ to $40°$. Attenuated total reflectance Fourier transform infrared (ATR-FTIR) spectra were acquired with a Nicolet iS50 spectrometer (Thermo Fisher Scientific, Waltham, MA, USA). The thermal behaviour of the samples was analysed using a Mettler Toledo TGA/DSC 3+ thermobalance (Mettler Toledo, Columbus, OH, USA), employing aluminium crucibles under a nitrogen flow of 50 mL per minute, across a temperature spanning from $25\ °C$ to $600\ °C$ at a heating rate of $10\ °C$ per minute. Analysis of these experiments was performed using Mettler Toledo STARe Evaluation Software version 16.10.

Single crystals of suitable quality of **[Mo$_2$O$_4$(MeOH)$_2$(L^1)]·2MeOH**, **[Mo$_2$O$_4$(H$_2$O)$_2$(L^2)]**, **[Mo$_2$O$_4$(MeOH)$_2$(L^3)]·2MeOH**, **[Mo$_2$O$_4$(MeOH)$_2$(L^4)]·2MeOH**, **[Mo$_2$O$_4$(MeOH)$_2$(L^5)]**, **[Mo$_2$O$_4$(MeOH)$_2$(L^6)]**, **[Mo$_2$O$_4$(MeOH)$_2$(L^7)]**, and **[Mo$_2$O$_4$(MeOH)$_2$(L^8)]·2MeOH** were obtained from diluted methanol solutions. Crystallographic data for **[Mo$_2$O$_4$(MeOH)$_2$(L^3)]·2MeOH** and **[Mo$_2$O$_4$(MeOH)$_2$(L^4)]·2MeOH** were obtained using ω-scans on an Oxford Xcalibur diffractometer with a 4-circle kappa goniometer and a CCD Sapphire 3 detector, using Mo Kα radiation at room temperature. The remaining compounds' data were gathered using a Rigaku XtaLAB Synergy-S diffractometer with a Dualflex source and a HyPix detector at 170(1) K. The data processing was carried out with the CrysAlis software suite [27]. Detailed crystal and intensity data collection and refinement parameters are presented in Supplementary Tables S1 and S2, and geometric data are reported in Tables S3–S7. Structure solution and refinement employed SHELXT [28] for dual space structure solution and SHELXL [29] for full-matrix least-squares refinement, treating non-hydrogen atoms anisotropically. Hydrogen atoms were modelled using geometrically idealized positions and the riding model, with their coordinates and distance constraints refined in later stages. The suite of SHELX programs was used within the Olex2 framework, [30] and Mercury 2021.3.0 software handled geometrical calculations and molecular graphics [31].

Nuclear Magnetic Resonance (NMR) spectroscopy was performed using a Bruker Avance III HD 400 MHz/54 mm Ascend spectrometer, equipped with a 5 mm PA BBI 1H/D BB Z-GRAD probehead. The range of experiments included 1D (^{1}H, ^{13}C-DEPTq) and 2D (COSY, ^{1}H–^{13}C HSQC, ^{1}H–^{13}C HMBC, ^{1}H–^{15}N HSQC, ^{1}H–^{15}N HMBC) techniques, conducted at room temperature. DMSO-$d6$ served as the solvent, with TMS as the internal standard for proton and carbon shifts, and nitrogen shifts using liquid ammonia as the standard.

In vitro biological evaluations were carried out to assess the cytotoxic effects of the complexes against cute monocytic leukaemia (THP-1) and hepatocellular carcinoma (HepG2) human cell lines, using the MTS assay as per established protocols [32]. The antibacterial efficacy of the complexes was tested against two Gram-positive bacteria (*Staphylococcus aureus* and *Enterococcus faecalis*) and two Gram-negative bacteria (*Escherichia coli* and *Moraxella catarrhalis*) using the broth microdilution method, adhering to CLSI guidelines [33,34].

3. Results and Discussion

3.1. Synthesis and Solid-State Characterization

Synthesis of the dioxomolybdenum(VI) complexes proceeded straightforwardly when reacting the methanolic solution of [MoO$_2$(acac)$_2$] with a suspension of ligands in methanol. The limiting factor of the reaction was the poor solubility of the ligands in methanol (and in any solvent other than DMSO, for that matter); thus, the reaction mixture had to be heated under reflux for two hours to guarantee reaction completion. Qualitatively, the solubility of the ligands in methanol can be described with the following scheme: H$_4$L^{3,7} ≈ H$_4$L^{4,8} > H$_4$L^{1,5} > H$_4$L^{2,6}. This trend was also reflected in the reaction times, i.e., ligands derived from dihydroxyaldehydes reacted faster than those derived from salicylaldehyde and 2-hydroxynaphthaldehyde. Owing to the overall low solubility of the resulting complexes, the isolated products were, in all cases, obtained as fine powder, and the crystals were only obtained by crystallization from very diluted solutions. In the case of Mo(VI) complexes with H$_4$L^2 and H$_4$L^4 ligands, it should be noted that reactions yielded [Mo$_2$O$_4$(MeOH)$_2$(L^2)] and [Mo$_2$O$_4$(MeOH)$_2$(L^4)], while the (re)crystallization of that material from diluted methanolic solutions gave single crystals of [Mo$_2$O$_4$(H$_2$O)$_2$(L^2)] and [Mo$_2$O$_4$(MeOH)$_2$(L^4)]·2MeOH, respectively. This is confirmed by a comparison of the PXRD data (Supplementary Materials, Figures S19–S26) with the simulated powder patterns obtained from crystal structure data (vide infra).

The thermal behavior of the prepared bulk complexes (Supplementary Materials, Figures S11–S18) is somewhat independent of the ligand choice and, expectedly, dependent on the presence of uncoordinated methanol molecules in the crystal. Interestingly, all methanol molecules (both coordinated and uncoordinated) are driven out of the bulk phase concomitantly, i.e., no separate steps are observed. As expected for molybdenum(VI) hydrazone complexes, at high temperatures, the materials undergo a complex sequence of decompositions, finally yielding MoO$_3$ above ~450 °C in an oxygen atmosphere. The mass fraction of the MoO$_3$ obtained through TGA analysis is consistent with the expected Mo:ligand ratio of 2:1.

Single-crystal X-ray diffraction: A detailed survey of the relevant literature revealed only few examples of the molybdenum(VI) complexes with ligands of the type [35–37]. All complexes investigated in this study are dinuclear, with the succinyl- and adipoyl-type ligands acting as ditopic hexadentate ones. The two compartments of each ligand coordinate in the hydrazonato form [38], after the deprotonation of O1−H1 and O2−H2 functionalities. In this way, the ligands behave as tetraanions, with their two compartments binding the two MoO$_2^{2+}$ units via O1, N1 and O2 donor atoms (Figures 1a,b and 2a,b; Supplementary Materials, Figures S1 and S2). The remaining coordination sites are occupied by the two oxido atoms, and an oxygen OH atom of ancillary methanol or, in the case of the [Mo$_2$O$_4$(H$_2$O)$_2$(L^2)] complex, water molecule (Figure 1b). Consequently, the coordination geometry of each Mo atom can be described as a distorted octahedral geometry, with the shortest distances being Mo=O ones, while those positioned trans to those are, expectedly, the longest within the coordination sphere. The Mo1 atom is, in all cases, shifted above the plane defined by the O1, N1, and O2 atoms towards the apical oxygen atom by 0.28–0.32 for the succinyl-type complexes, whereas for the adipoyl ones these distances are larger, between ca 0.31 and 0.36.

When considering the molecular structures of dinuclear complexes based on the succinyl-type of ligands, one observes that, except [Mo$_2$O$_4$(H$_2$O)$_2$(L^2)], the remaining ones are conformationally fairly similar (Figure 1c). Namely, conformations of complexes are staggered such that the planes defined by the aldehyde residues are parallel and distanced by ca 0.66, 0.77, and 0.80 for [Mo$_2$O$_4$(MeOH)$_2$(L^4)], [Mo$_2$O$_4$(MeOH)$_2$(L^3)], and [Mo$_2$O$_4$(MeOH)$_2$(L^1)], respectively, whereas in the case of [Mo$_2$O$_4$(H$_2$O)$_2$(L^2)], this distance is ca 0.34. Due to the longer bridge between the two compartments of the complex, conformational differences in the group of adipoyl-type complexes are larger (Figure 2c). Namely, the molecule of [Mo$_2$O$_4$(MeOH)$_2$(L^8)] is essentially planar, while [Mo$_2$O$_4$(MeOH)$_2$(L^5)], and [Mo$_2$O$_4$(MeOH)$_2$(L^7)] adopt staggered conformations. Moreover, in [Mo$_2$O$_4$(MeOH)$_2$(L^6)] and [Mo$_2$O$_4$(MeOH)$_2$(L^7)], planes defined by the aldehyde residues are not parallel but form angles of ca 32° and ca 26°, respectively.

In solvates of the succinyl-type of complexes, [Mo$_2$O$_4$(MeOH)$_2$(L^1)]·2MeOH, textbf[Mo$_2$O$_4$(MeOH)$_2$(L^3)]·2MeOH, and [Mo$_2$O$_4$(MeOH)$_2$(L^4)]·2MeOH, non-coordinated methanol molecules bridge coordination entities by acting both as hydrogen bond donors and acceptors (Figures 3a and S3–S5). The situation is different in [Mo$_2$O$_4$(H$_2$O)$_2$(L^2)], where the coordinated water molecules act solely as hydrogen bond donors to connect neighbouring complexes via O–H···N and O–H···O hydrogen bonds (Figures 3b and S6). The resulting supramolecular architectures are, in all cases, further supported and stabilized by a plethora of C–H···O hydrogen bonds. Similarly, in the crystal structures of adipoyl-type complexes, supramolecular interactions are mediated through the only available hydrogen bond donors, the coordinated methanol OH and aryl hydroxy OH group, while hydrogen bond acceptors are either imide nitrogen atom or the axial {MoO$_2$}$^{2+}$ oxygen atom (Figures 4 and S7–S10). The presence of non-coordinated methanol molecules in the dual role of hydrogen bond donors and acceptors in [Mo$_2$O$_4$(MeOH)$_2$(L^8)]·2MeOH makes the hydrogen-bonded network comparatively richer (Figures 4b and S10). Finally, as in the case of succinyl-type complexes, crystal structures in adipoyl-type of complexes are stabilized by a collection of C–H···O hydrogen bonds.

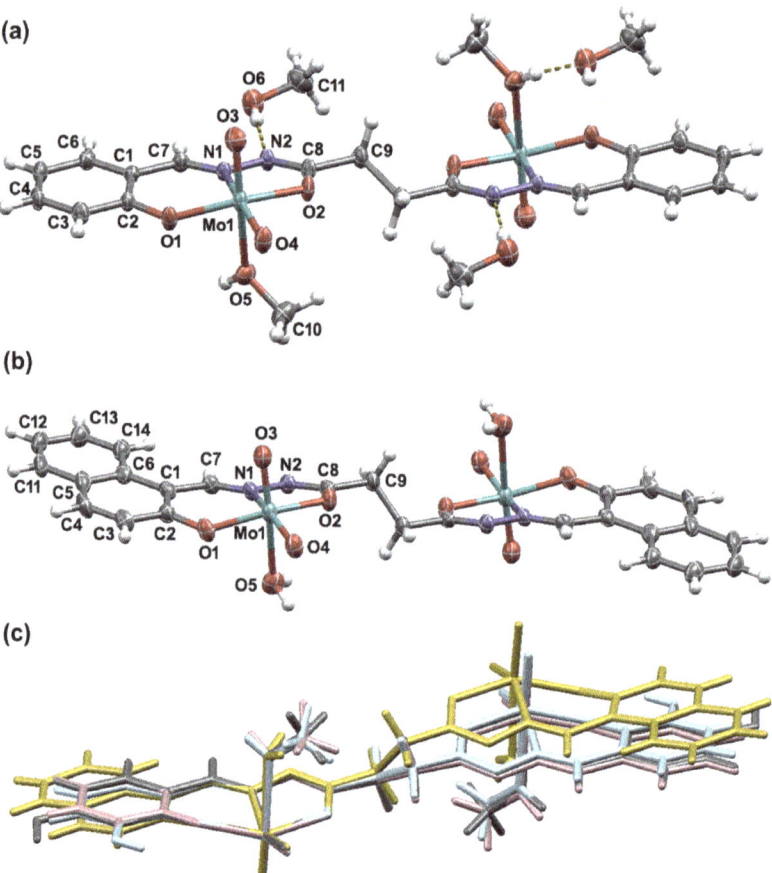

Figure 1. Molecular structures of: (**a**) [Mo$_2$O$_4$(MeOH)$_2$(L^1)]·2MeOH and (**b**) [Mo$_2$O$_4$(H$_2$O)$_2$(L^2)] with the atom numbering schemes. The central aliphatic C2 fragments of molecules lie on the inversion center. (**c**) Overlay of the molecular structures of [Mo$_2$O$_4$(MeOH)$_2$(L^1)] (pink), [Mo$_2$O$_4$(H$_2$O)$_2$(L^2)] (yellow), [Mo$_2$O$_4$(MeOH)$_2$(L^3)] (blue), and [Mo$_2$O$_4$(MeOH)$_2$(L^4)] (gray). In (**a**,**b**) displacement ellipsoids are drawn at the 50% probability level, while the hydrogen atoms are presented as spheres of arbitrary small radii. In (**a**), hydrogen bonds are highlighted by yellow dashed lines.

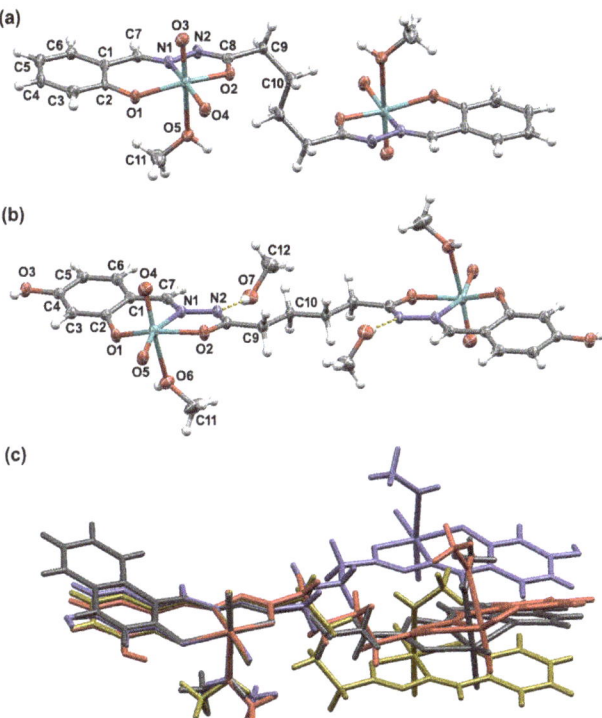

Figure 2. Molecular structures of: (**a**) [Mo$_2$O$_4$(MeOH)$_2$(L^5)] and (**b**) [Mo$_2$O$_4$(MeOH)$_2$(L^8)]·2MeOH with the atom numbering schemes. The central aliphatic C4 fragments of molecules sit on the inversion center. (**c**) Overlay of the molecular structures of [Mo$_2$O$_4$(MeOH)$_2$(L^5)] (yellow), [Mo$_2$O$_4$(H$_2$O)$_2$(L^6)] (grey), [Mo$_2$O$_4$(MeOH)$_2$(L^7)] (red), and [Mo$_2$O$_4$(MeOH)$_2$(L^8)] (blue). In (**a**,**b**) displacement ellipsoids are drawn at the 50% probability level, while the hydrogen atoms are presented as spheres of arbitrary small radii. In (**a**), hydrogen bonds are highlighted by yellow dashed lines.

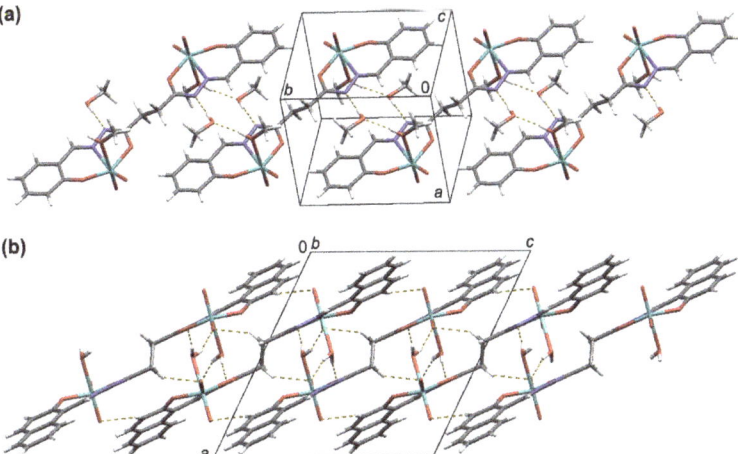

Figure 3. Supramolecular architectures found in: (**a**) [Mo$_2$O$_4$(MeOH)$_2$(L^1)]·2MeOH and (**b**) [Mo$_2$O$_4$(H$_2$O)$_2$(L^2)]. Hydrogen bonds are highlighted as yellow dashed lines.

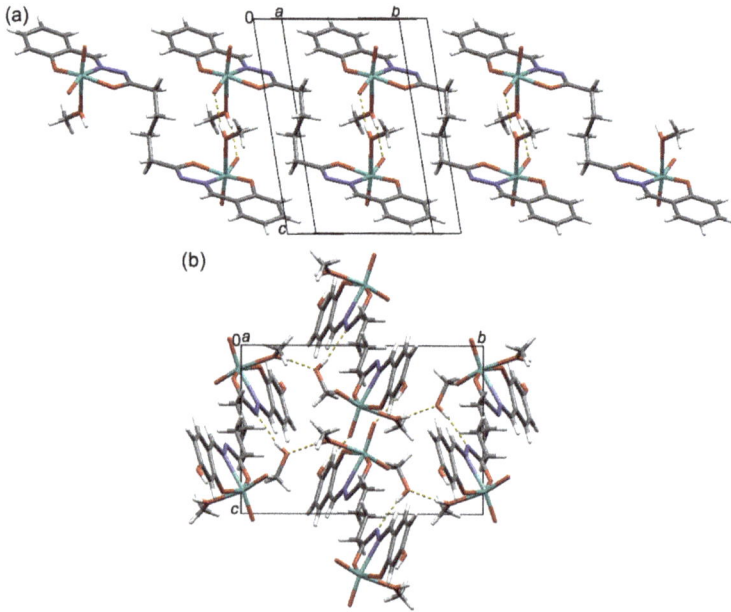

Figure 4. Crystal packing in: (a) [Mo$_2$O$_4$(MeOH)$_2$(L^5)] and (b) [Mo$_2$O$_4$(MeOH)$_2$(L^8)]·2MeOH. Hydrogen bonds are highlighted as yellow dashed lines.

3.2. NMR and FT-IR Spectroscopy

The chemical identities of the title complexes in DMSO-d_6 solutions at room temperature were established via ^{1}H, ^{13}C, and ^{15}N NMR spectroscopy (Supplementary Materials, Tables S8 and S9, Figures S27–S32). Relevant chemical shifts unveil that, in the DMSO solution, the explored dinuclear Mo(VI) complexes are symmetrical, since only one set of signals was observed in each case. Moreover, these complexes, unlike the related neutral ligands, in DMSO-d_6 solutions, dominantly adopt one isomeric form. The absence of hydroxyaryl O1–H1 and hydrazonic N2–H2 protons in the spectra of complexes confirms that coordinated ligands are present in their tetraanionic form. Moreover, the significant deshielding of the N2 nitrogen chemical shift in the complexes, in comparison to neutral ligands, along with the slight deshielding of the C8 signal, confirms the enol-imino form of the coordinated ligands [39]. The ^{1}H, ^{13}C, and ^{15}N chemical shifts in complexes are comparable to those of neutral ligands. Finally, owing to the strong donor nature of the DMSO, the title complexes undergo a solvent exchange reaction, which has been previously described in the literature for a similar type of Mo complex [40].

FTIR-ATR spectra (Supplementary Materials, Figures S33–S40) show features characteristic of dioxomolybdenum(VI) hydrazone complexes. In contrast to the respective ligands [15], absorption peaks corresponding to C=O stretching vibrations (usually found at ~1650 cm^{-1}) and those related to N-H stretching vibrations (at around ~3200 cm^{-1}) are absent in the spectra of prepared complexes. Thus, it can be concluded that the ligand is (tetra)deprotonated and in enol-imino form. However, the high-wave-number region of the spectra is rich, owing to the O-H stretching vibrations of crystal and coordinated methanol molecules, and in the case of complexes derived from H$_4$L^3, H$_4$L^4, H$_4$L^7, and H$_4$L^8, additional phenolic O-H bands. A significant shift to lower wavenumbers of the C=N stretching bands (~1590–1600 cm^{-1}) compared to those of the free ligands (~1610–1630 cm^{-1}), together with the absence of the C=O band, corroborates the coordination of the {MoO$_2$}$^{2+}$ core by the *ONO* pincer-like compartment of the ligand in its hydrazonato form. The C-O stretching vibrations of the enolato and phenolato fragments can be found at ~1330 and ~1270 cm^{-1}, respectively. All complexes exhibit distinct bands at ~930–940 cm^{-1} and

~880–910 cm^{-1}, typical for dioxomolybdenum(VI) complexes, corresponding to symmetric and asymmetric {MoO$_2$}$^{2+}$ core stretching vibrations, respectively [41]. Moreover, the spectra display bands indicative of coordinated methanol molecules, i.e., C–O stretching vibrations at ca. 1010–1030 cm^{-1}. Taken together, the spectroscopic data are consistent with the identities and structures of the prepared complexes obtained from SC-XRD experiments.

3.3. In Vitro Cytotoxic and Antibacterial Activity

The here-reported molybdenum(VI) complexes were tested for their cytotoxic activity against HepG2 and THP-1 cells, while their antibacterial activity was evaluated *on S. aureus, E. faecalis, E. coli*, and *M. catarrhalis* bacterial strains. The bioassay results are summarized in Table 1.

Table 1. The IC$_{50}$ values and minimum inhibitory concentrations (MIC) of the Mo(VI) complexes.

Compound	IC$_{50}$ (µmol L^{-1})		MIC (µg mL^{-1})			
	THP-1	HepG2	S. aureus	E. faecalis	E. coli	M. catarrhalis
1-Mo	19.58	>100	>256	64	>256	32
2-Mo	7.04	>100	32	32	32	4
3-Mo	>100	>100	256	256	128	128
4-Mo	>100	>100	>256	128	>256	>256
5-Mo	>100	>100	>256	>256	>256	>256
6-Mo	9.78	57.98	>256	128	128	2
7-Mo	>100	>100	>256	32	64	64
8-Mo	>100	>100	>256	>256	>256	>256
staurosporine	0.10	7.98	–	–	–	–
azithromycin	–	–	1	8	0.50	0.125

1-Mo = [Mo$_2$O$_4$(MeOH)$_2$(L^1)]; 2-Mo = [Mo$_2$O$_4$(MeOH)$_2$(L^2)]; 3-Mo = [Mo$_2$O$_4$(MeOH)$_2$(L^3)]; 4-Mo = [Mo$_2$O$_4$(MeOH)$_2$(L^4)]; 5-Mo = [Mo$_2$O$_4$(MeOH)$_2$(L^5)]; 6-Mo = [Mo$_2$O$_4$(MeOH)$_2$(L^6)]; 7-Mo = [Mo$_2$O$_4$(MeOH)$_2$(L^7)]; and 8-Mo = [Mo$_2$O$_4$(MeOH)$_2$(L^8)].

According to the results (Table 1), all of the investigated complexes were found to be non-cytotoxic towards HepG2 cells, and, except **[Mo$_2$O$_4$(MeOH)$_2$(L^1)]**, **[Mo$_2$O$_4$(MeOH)$_2$(L^2)]** and **[Mo$_2$O$_4$(MeOH)$_2$(L^6)]**, which displayed only weak to moderate cytotoxicity, did not show any cytotoxic effects against THP-1. The results, given in Table 1, also reveal that complexes in general exhibited poor or no antibacterial activity, while some of them with MICs equal to 32 µg mL^{-1} demonstrated mild activity. Only **[Mo$_2$O$_4$(MeOH)$_2$(L^2)]** and **[Mo$_2$O$_4$(MeOH)$_2$(L^6)]**, with minimum inhibitory concentrations equal to 4 and 2 µg mL^{-1}, respectively, were proved to possess appreciable antibacterial potential towards *M. catarrhalis*.

Apart from **[Mo$_2$O$_4$(MeOH)$_2$(L^1)]**, **[Mo$_2$O$_4$(MeOH)$_2$(L^2)]**, and **[Mo$_2$O$_4$(MeOH)$_2$(L^6)]**, chelation in general did not have a large influence on the cytotoxic properties of the examined compounds [15]. However, some differences in the antibacterial properties of Mo(VI) complexes in comparison with the corresponding free ligands were established. On the one hand, a noticeable reduction in antibacterial properties after the complexation to molybdenum was only observed for the anti-*E. faecalis* activity of H$_4$L^5, as well as for the anti-*E. faecalis* and anti-*E. coli* activities of H$_4$L^6. On the other hand, the bactericidal potency of **[Mo$_2$O$_4$(MeOH)$_2$(L^2)]** and **[Mo$_2$O$_4$(MeOH)$_2$(L^7)]** complexes was demonstrated, along with the increased anti-*M. catarrhalis* activity of **[Mo$_2$O$_4$(MeOH)$_2$(L^6)]**, when compared to the neutral ligands.

4. Conclusions

A series of novel molybdenum(VI) complexes with succinyl and adipoyl dihydrazones was synthesized, characterized, and evaluated with respect to their biological activity. Conventional solution synthesis proved to be a straightforward route towards the title complexes, although it was to some extent limited by the solubility of the ligands. The structures of the complexes were explored via single-crystal X-ray diffraction and FTIR spectroscopy in the solid state, while their behavior in the DMSO-d_6 solution was studied by the NMR technique. Structural studies unveiled that all complexes are dinuclear, with tetraanions of dihydrazones acting as flexible ditopic hexadentate ligands, both in the solid state and in DMSO solution. The complexes generally showed non-cytotoxic behavior towards HepG2 cells and varying degrees of cytotoxicity against THP-1 cells. Some exhibited mild antibacterial activity, with notable effectiveness against *Moraxella catarrhalis*. The results highlight the moderate improvement in antimicrobial properties compared to the respective ligands. These findings, combined with the facile modulation of the complexes of this type, could serve as a fruitful platform for the future development of this type of compound as potentially bioactive compounds.

Supplementary Materials: The following supporting information can be downloaded at: https://www.mdpi.com/article/10.3390/cryst14020135/s1, Scheme S1: Reaction scheme; Tables S1 and S2: General and crystallographic data; Tables S3–S7: Selected geometric data; Figures S1–S10: Crystal structure representations; Figures S11–S18: TGA/DSC data; Figures S19–S26: Comparison of PXRD patterns; Figures S27–S32: NMR spectra Tables S8 and S9: The NMR numbering schemes; ^{1}H and ^{13}C and ^{15}N assignments; Figures S33–S40: ATR FT-IR spectra.

Author Contributions: Investigation, formal analysis, writing—original draft preparation, E.T., V.D., K.P. and M.R.; visualization, E.T. and M.R.; writing—review and editing, E.T., V.D., K.P. and M.R. All authors have read and agreed to the published version of the manuscript.

Funding: This work has been fully supported by Croatian Science Foundation under the project (IP-2016-06-4221). We acknowledge the support of project CIuK co-financed by the Croatian Government and the European Union through the European Regional Development Fund–Competitiveness and Cohesion Operational Programme (Grant KK.01.1.1.02.0016).

Data Availability Statement: Crystallographic data sets for the structures [$Mo_2O_4(MeOH)_2(L^1)$]·2MeOH, [$Mo_2O_4(H_2O)_2(L^2)$], [$Mo_2O_4(MeOH)_2(L^3)$]·2MeOH, [$Mo_2O_4(MeOH)_2(L^4)$]·2MeOH, [$Mo_2O_4(MeOH)_2(L^5)$], [$Mo_2O_4(MeOH)_2(L^6)$], [$Mo_2O_4(MeOH)_2(L^7)$] and [$Mo_2O_4(MeOH)_2(L^8)$]·2MeOH are available through the Cambridge Structural Database with deposition numbers CCDC 2314177-2314184. These data can be obtained free of charge via https://www.ccdc.cam.ac.uk/structures/ (accessed on 26 January 2024).

Acknowledgments: We are grateful to Ljubica Ljubić, Kristina Prezelj, for the assistance in the synthetic procedures and Nikola Cindro for the help with the NMR measurements.

Conflicts of Interest: The authors declare no conflicts of interest.

References

1. Mali, S.N.; Thorat, B.R.; Gupta, D.R.; Pandey, A. Mini-Review of the Importance of Hydrazides and Their Derivatives—Synthesis and Biological Activity. *Eng. Proc.* **2021**, *11*, 21.
2. Tatum, L.A.; Su, X.; Aprahamian, I. Simple Hydrazone Building Blocks for Complicated Functional Materials. *Acc. Chem. Res.* **2014**, *47*, 2141–2149. [CrossRef] [PubMed]
3. Uribe-Romo, F.J.; Doonan, C.J.; Furukawa, H.; Oisaki, K.; Yaghi, O.M. Crystalline covalent organic frameworks with hydrazone linkages. *J. Am. Chem. Soc.* **2011**, *133*, 11478–11481. [CrossRef] [PubMed]
4. Hossain, M.K.; Plutenko, M.O.; Schachner, J.A.; Haukka, M.; Mösch-Zanetti, N.C.; Fritsky, I.O. Dioxomolybdenum(VI) complexes of hydrazone phenolate ligands—Syntheses and activities in catalytic oxidation reactions. *J. Indian Chem. Soc.* **2021**, *98*, 100006. [CrossRef]
5. Guskos, N.; Likodimos, V.; Glenis, S.; Typek, J.; Wabia, M.; Paschalidis, D.G.; Tossidis, I.; Lin, C.L. Magnetic properties of rare-earth hydrazone compounds. *J. Magn. Magn. Mater.* **2004**, *272*, 1067–1069. [CrossRef]
6. Liu, R.; Cui, J.; Ding, T.; Liu, Y.; Liang, H. Research Progress on the Biological Activities of Metal Complexes Bearing Polycyclic Aromatic Hydrazones. *Molecules* **2022**, *27*, 8393. [CrossRef] [PubMed]
7. Tupolova, Y.P.; Popov, L.D.; Vlasenko, V.G.; Gishko, K.B.; Kapustina, A.A.; Berejnaya, A.G.; Golubeva, Y.A.; Klyushova, L.S.; Lider, E.V.; Lazarenko, V.A.; et al. Crystal structure and cytotoxic activity of Cu(ii) complexes with bis-benzoxazolylhydrazone of 2,6-diacetylpyridine. *New J. Chem.* **2023**, *47*, 14972–14985. [CrossRef]
8. Tupolova, Y.P.; Shcherbakov, I.N.; Popov, L.D.; Vlasenko, V.G.; Gishko, K.B.; Kapustina, A.A.; Berezhnaya, A.G.; Golubeva, Y.A.; Klyushova, L.S.; Lider, E.V.; et al. Copper coordination compounds based on bis-quinolylhydrazone of 2,6-diacetylpyridine: Synthesis, structure and cytotoxic activity. *Polyhedron* **2023**, *233*, 116292. [CrossRef]
9. Su, X.; Aprahamian, I. Hydrazone-based switches, metallo-assemblies and sensors. *Chem. Soc. Rev.* **2014**, *43*, 1963–1981. [CrossRef]

10. Zavalishin, M.N.; Gamov, G.A.; Pimenov, O.A.; Pogonin, A.E.; Aleksandriiskii, V.V.; Usoltsev, S.D.; Marfin, Y.S. Pyridoxal 5′-phosphate 2-methyl-3-furoylhydrazone as a selective sensor for Zn^{2+} ions in water and drug samples. *J. Photochem. Photobiol. A* **2022**, *432*, 114112. [CrossRef]
11. Bagherian, N.; Karimi, A.R.; Amini, A. Chemically stable porous crystalline macromolecule hydrazone-linked covalent organic framework for CO_2 capture. *Colloids Surf. A* **2021**, *613*, 126078. [CrossRef]
12. Golla, U.; Adhikary, A.; Mondal, A.K.; Tomar, R.S.; Konar, S. Synthesis, structure, magnetic and biological activity studies of bis-hydrazone derived Cu(ii) and Co(ii) coordination compounds. *Dalton Trans.* **2016**, *45*, 11849–11863. [CrossRef]
13. Wang, M.; Cheng, C.; Chunbo, L.; Wu, D.; Song, J.; Wang, J.; Zhou, X.; Xiang, H.; Liu, J. Smart, chiral, and non-conjugated cyclohexane-based bissalicylaldehyde hydrazides: Multi-stimuli-responsive, turn-on, ratiometric, and thermochromic fluorescence, single crystal structures, and DFT calculations. *J. Mater. Chem. C* **2019**, *7*, 6767–6778. [CrossRef]
14. Chen, Z.; Zhou, S.; Shen, Y.; Zou, H.; Liu, D.; Liang, F. Copper(II) Clusters of Two Pairs of 2,3-Dihydroxybutanedioyl Dihydrazones: Synthesis, Structure, and Magnetic Properties. *Eur. J. Inorg. Chem.* **2014**, *2014*, 5783–5792. [CrossRef]
15. Chen, Z.; Shen, Y.; Li, L.; Zou, H.; Fu, X.; Liu, Z.; Wang, K.; Liang, F. High-nuclearity heterometallic clusters with both an anion and a cation sandwiched by planar cluster units: Synthesis, structure and properties. *Dalton Trans.* **2017**, *46*, 15032–15039. [CrossRef] [PubMed]
16. Mondal, A.K.; Jena, H.S.; Malviya, A.; Konar, S. Lanthanide-Directed Fabrication of Four Tetranuclear Quadruple Stranded Helicates Showing Magnetic Refrigeration and Slow Magnetic Relaxation. *Inorg. Chem.* **2016**, *55*, 5237–5244. [CrossRef] [PubMed]
17. Schattschneider, C.; Doniz Kettenmann, S.; Hinojosa, S.; Heinrich, J.; Kulak, N. Biological activity of amphiphilic metal complexes. *Coord. Chem. Rev.* **2019**, *385*, 191–207. [CrossRef]
18. Liang, J.; Sun, D.; Yang, Y.; Li, M.; Li, H.; Chen, L. Discovery of metal-based complexes as promising antimicrobial agents. *Eur. J. Med. Chem.* **2021**, *224*, 113696. [CrossRef]
19. Verma, G.; Marella, A.; Shaquiquzzaman, M.; Akhtar, M.; Ali, M.R.; Alam, M.M. A review exploring biological activities of hydrazones. *J. Pharm. Bioallied Sci.* **2014**, *6*, 69–80.
20. Le Goff, G.; Ouazzani, J. Natural hydrazine-containing compounds: Biosynthesis, isolation, biological activities and synthesis. *Bioorg. Med. Chem.* **2014**, *22*, 6529–6544. [CrossRef]
21. Kumar, P.; Narasimhan, B. Hydrazides/hydrazones as antimicrobial and anticancer agents in the new millennium. *Mini Rev. Med. Chem.* **2013**, *13*, 971–987. [CrossRef] [PubMed]
22. Popiołek, Ł. Hydrazide-hydrazones as potential antimicrobial agents: Overview of the literature since 2010. *Med. Chem. Res.* **2017**, *26*, 287–301. [CrossRef]
23. Arora, T.; Devi, J.; Boora, A.; Taxak, B.; Rani, S. Synthesis and characterization of hydrazones and their transition metal complexes: Antimicrobial, antituberculosis and antioxidant activity. *Res. Chem. Intermed.* **2023**, *49*, 4819–4843. [CrossRef]
24. Ullah, H.; Previtali, V.; Mihigo, H.B.; Twamley, B.; Rauf, M.K.; Javed, F.; Waseem, A.; Baker, R.J.; Rozas, I. Structure-activity relationships of new Organotin(IV) anticancer agents and their cytotoxicity profile on HL-60, MCF-7 and HeLa human cancer cell lines. *Eur. J. Med. Chem.* **2019**, *181*, 111544. [CrossRef] [PubMed]
25. Topić, E.; Damjanović, V.; Pičuljan, K.; Vrdoljak, V.; Rubčić, M. Succinyl and Adipoyl Dihydrazones: A Solid-State, Solution and Antibacterial Study. *Crystals* **2022**, *12*, 1175. [CrossRef]
26. Gherke, H., Jr.; Veal, J. Acetylacetonate complexes of molybdenum (V) and molybdenum (VI). I. *Inorg. Chim. Acta* **1969**, *3*, 623–627. [CrossRef]
27. *CrysAlisPro Software System*, version 1.171.41.92a, Rigaku Oxford Diffraction: Oxford, UK, 2020.
28. Sheldrick, G.M. SHELXT-Integrated space-group and crystal-structure determination. *Acta Crystallogr. A* **2015**, *71*, 3–8. [CrossRef]
29. Sheldrick, G.M. Crystal structure refinement with SHELXL. *Acta Crystallogr. C* **2015**, *71*, 3–8. [CrossRef]
30. Dolomanov, O.V.; Bourhis, L.J.; Gildea, R.J.; Howard, J.A.K.; Puschmann, H. OLEX2: A complete structure solution, refinement and analysis program. *J. Appl. Crystallogr.* **2009**, *42*, 339–341. [CrossRef]
31. Macrae, C.F.; Sovago, I.; Cottrell, S.J.; Galek, P.T.A.; McCabe, P.; Pidcock, E.; Platings, M.; Shields, G.P.; Stevens, J.S.; Towler, M.; et al. Mercury 4.0: From visualization to analysis, design and prediction. *J. Appl. Crystallogr.* **2020**, *53*, 226–235. [CrossRef]
32. Mosmann, T. Rapid colorimetric assay for cellular growth and survival: Application to proliferation and cytotoxicity assays. *J. Immunol. Methods* **1983**, *65*, 55–63. [CrossRef] [PubMed]
33. Rubčić, M.; Pisk, J.; Pičuljan, K.; Damjanović, V.; Lovrić, J.; Vrdoljak, V. Symmetrical disubstituted carbohydrazides: From solid-state structures to cytotoxic and antibacterial activity. *J. Mol. Struct.* **2019**, *1178*, 222–228. [CrossRef]
34. *CLSI Document M07-A8*; Methods for Dilution Antimicrobial Susceptibility Tests for Bacteria That Grow Aerobically. Clinical and Laboratory Standards Institute: Wayne, PA, USA, 2009.
35. Ngan, N.K.; Lo, K.M.; Wong, C.S.R. Dinuclear and polynuclear dioxomolybdenum(VI) Schiff base complexes: Synthesis, structural elucidation, spectroscopic characterization, electrochemistry and catalytic property. *Polyhedron* **2012**, *33*, 235. [CrossRef]
36. Kurbah, S.D.; Kumar, A.; Shangpung, S.; Syiemlieh, I.; Khongjoh, I.; Lal, R.A. Synthesis, Characterization, and Fluorescence Chemosensor Properties of a cis-Dioxomolybdenum(VI) Complex Containing Multidentate Hydrazone Ligands. *Z. Anorg. Allg. Chem.* **2017**, *643*, 794–801. [CrossRef]
37. Kurbah, S.D.; Kumar, A.; Syiemlieh, I.; Asthana, M.; Lal, R.A. Bimetallic cis-dioxomolybdenum(VI) complex containing hydrazone ligand: Syntheses, crystal structure and catalytic studies. *Inorg. Chem. Commun.* **2017**, *86*, 39–43. [CrossRef]

38. Vrdoljak, V.; Hrenar, T.; Rubčić, M.; Pavlović, G.; Friganović, T.; Cindrić, M. Ligand-Modulated Nuclearity and Geometry in Nickel(II) Hydrazone Complexes: From Mononuclear Complexes to Acetato- and/or Phenoxido-Bridged Clusters. *Int. J. Mol. Sci.* **2023**, *24*, 1909. [CrossRef]
39. Maurya, M.R.; Saini, N.; Avecilla, F. Effect of N-based additive on the optimization of liquid phase oxidation of bicyclic, cyclic and aromatic alcohols catalyzed by dioxidomolybdenum(vi) and oxidoperoxidomolybdenum(vi) complexes. *RSC Adv.* **2015**, *5*, 101076. [CrossRef]
40. Pisk, J.; Rubčić, M.; Kuzman, D.; Cindrić, M.; Agustin, D.; Vrdoljak, V. Molybdenum(vi) complexes of hemilabile aroylhydrazone ligands as efficient catalysts for greener cyclooctene epoxidation: An experimental and theoretical approach. *New J. Chem.* **2019**, *43*, 5531–5542. [CrossRef]
41. Berg, J.M.; Holm, R.H. Structure proofs of ligated and polymeric dioxomolybdenum(VI)-tridentate complexes: $MoO_2(C_5H_3N-2,6-(CH_2S)_2)(C_4H_8SO)$ and $[MoO_2(C_5H_3N-2,6-(CH_2O)_2)]n$. *Inorg. Chem.* **1983**, *22*, 1768–1771. [CrossRef]

Disclaimer/Publisher's Note: The statements, opinions and data contained in all publications are solely those of the individual author(s) and contributor(s) and not of MDPI and/or the editor(s). MDPI and/or the editor(s) disclaim responsibility for any injury to people or property resulting from any ideas, methods, instructions or products referred to in the content.

MDPI
St. Alban-Anlage 66
4052 Basel
Switzerland
www.mdpi.com

Crystals Editorial Office
E-mail: crystals@mdpi.com
www.mdpi.com/journal/crystals

Disclaimer/Publisher's Note: The statements, opinions and data contained in all publications are solely those of the individual author(s) and contributor(s) and not of MDPI and/or the editor(s). MDPI and/or the editor(s) disclaim responsibility for any injury to people or property resulting from any ideas, methods, instructions or products referred to in the content.

www.ingramcontent.com/pod-product-compliance
Lightning Source LLC
LaVergne TN
LVHW070603100526
838202LV00012B/552